Focus
Forecasting

Focus Forecasting

Computer Techniques for Inventory Control

Bernard T. Smith
Foreword by Oliver W. Wight

THE *Oliver Wight* ® COMPANIES

Oliver Wight Limited Publications, Inc.
5 Oliver Wight Drive
Essex Junction, VT 05452

Library of Congress Cataloging in Publication Data

Smith, Bernard T   1938-
Focus forecasting : computer techniques for
inventory control.

Includes index.
1. Inventory control–Data processing.
2. Forecasting–Mathematical
models.   I. Title.
HD55.S48   658.7′87′02854   78-6120
ISBN: 0-939246-02-3

Manufactured in the United States of America by
The Maple-Vail Book Manufacturing Group

3 4 5 6 7 8 9 10

Contents

Foreword

My field is production and inventory management. The last few years have seen a revolution in our thinking on manufacturing inventory management. We now see it as a by-product of scheduling. MRP (material requirements planning), originally invented to order parts for a product, has developed into a network scheduling technique at the heart of a system that includes master scheduling, capacity planning, shop scheduling, and vendor scheduling.

Distribution inventory management hasn't made the same kind of progress. Virtually everything I have read is just a rehash of the same old stuff. From my work with the IBM IMPACT programs, I know the limitations of these statistical inventory control concepts.

When Bernie Smith, whom I've known for many years, came to me with the idea for a book, I was skeptical. Most books are long on theory and short on the practical approach. But when I saw what Bernie was doing, I became interested. The underlying philosophy of focus forecasting is sound and follows the same type of development that we've seen in manufacturing. His approach shies away from "optimization," where the computer comes up with the one right answer, and instead emphasizes *simulation.*

Focus forecasting tests many different simple forecasting strategies *each time* inventory is to be ordered. Then it chooses the strategy that would have worked best the last time around. It's well known that the immediate past is usually the best indicator of the immediate future, but Bernie has taken this a step further. He uses the computer to simulate and chooses the forecasting *technique* that would have forecast the immediate past best and uses that technique to forecast the immediate future.

Another basic tenet of Bernie's approach is that the system doesn't do it, the people do. Most of us who have worked in the field have had to learn the hard way that techniques accomplish nothing. They are just tools for people.

From some rough ideas to an original manuscript, a second revision, and a final third revision, Bernie's book has grown into one that I'm proud to have played a small part in. I believe this is the first fresh thinking about distribution inventory management that we've seen in the last twenty years. I suspect that some theoreticians will be dismayed at the simplicity of Bernie's approaches. I am not. Only results matter, and these approaches to distribution inventory management have produced more tangible results than any other technique I've ever seen.

Oliver W. Wight
Blodgett Landing, New Hampshire

Preface

This is a simple book. It is based on my fourteen years of learning about inventory management. It's simple because I've learned over the years that complex answers don't work. As a matter of fact I believe that a complex answer is just used to hide the fact that the answer has not yet been found. Unfortunately it took me fourteen years to learn the complex answers and finally settle on the simple answers.

My formal training is in accounting. I taught accounting at the graduate level in the University of Bridgeport in Connecticut. From accounting I learned about the profit impact of inventory management decisions. When I write about conflicts between the finance manager and the inventory manager I'm trying to represent both views objectively. I've experienced the frustrations of cash budgets overburdened by inventory excesses, and I've listened to customer service complaints caused by panic inventory cutbacks to generate cash.

This book is about a revolutionary new forecasting concept called focus forecasting. It's revolutionary because of its consistent accuracy and its basic simplicity. During any three-month period in the past twelve months, the total dollars that focus forecasting predicted and the actual demand that occurred differed by less than 1 percent. During the energy crisis of 1973 focus forecasting predicted demand within 10 percent. During the economic recession that immediately followed the energy crisis in 1974 focus forecasting was still predicting demand within 10 percent. In a stable economy focus forecasting predicts demand with uncanny accuracy. During April, May, and June 1976, the difference between the actual demand in dollars and the demand that focus forecasting predicted was only 0.3 percent.

This book isn't just the story of a revolutionary forecasting concept; it also covers a finished goods inventory management system and the people who make that system work. The focus forecasting inventory management system buys over 100,000 stock-keeping units every month. It suggests buying decisions for every item. The buyers who review those suggestions change less than 8 percent of them. At the end of the month the dollar total of all suggested buys differs from the actual purchase order dollars by less than 5 percent.

This book is about a revolutionary forecasting approach and an inventory management system. But more, it's about the people who helped create that system and the people who use it to make the everyday inventory management decisions. The magic of focus forecasting is that it allows people to make decisions. It allows them to understand the tools the computer is using and it holds them accountable for inventory level and customer service.

The concepts in this book are based on three principles: (1) People will only use simple systems. (2) Simple systems are the only systems that work. (3) The only measure of any system is how well it works.

This book is not a cookbook. You will not be able to have your programmers put focus forecasting techniques into your computer to run your inventory management systems. The book will tell you about some new approaches to forecasting and inventory management. It will remind you of the principles that guide the development of all new systems.

Focus forecasting is not a formula. Focus forecasting uses a series of simple forecasting approaches. It uses the powerful simulative ability of the computer to pick the one forecasting strategy that will work best for this item for this time. This forecasting system is transparent to the user. The user knows the simple strategy that was chosen and why it was chosen.

Focus forecasting wasn't possible ten years ago. Ten years ago computers didn't run fast enough to use this approach. Somebody once said there's nothing new under the sun. Indeed, in one sense, focus forecasting is not new. It is firmly rooted in the systems and concepts developed for early computers. It's only new in that it is the next logical step in developing a computer forecasting system.

One of the earliest forecasting concepts was the "recipe approach," the foundation of all forecasts including focus forecasting. The recipe approach explains demand in terms of trends, cycles, seasons, and noise. Trends are the momentum of demand. Is the demand going up? Is the demand going down? Cycles are long-term demand patterns that repeat themselves every four or five years. These demand patterns are closely related to business cycles. Seasons are demand patterns that repeat themselves every year because of their relationship to climatic changes or other special annual events. Noise is the part of demand that the recipe approach cannot explain. It is the random distortion of a seemingly predictable demand pattern. Focus forecasting uses concepts developed by the recipe approach. It doesn't call them trends, cycles, seasons, and noise, but it takes account of all four of them.

Exponential smoothing was an early application of the recipe approach to computer forecasting. In its simplest form exponential smoothing is an approximation of a moving average. Later versions of exponential smoothing incorporated trends and seasonality. Inventory management systems based on exponential smoothing use the concept of noise to set safety stocks. There are many different kinds of exponential smoothing now that use double smoothing, triple smoothing, and adaptive smoothing. Still, exponential smoothing is most widely used in its original form—an approximation of a moving average. It was because of exponential smoothing and the need for better inventory management systems that people went through the discipline of gathering demand data and putting it on the computer. Exponential smoothing was the catalyst for founding a data base for all inventory management systems in use today. In my company focus forecasting uses a data base created by exponential smoothing.

The moving averages used in exponential smoothing can cover different time periods—two weeks, two months, three months, eight months, a year, two years. People wondered what moving average period would best forecast the future. IBM and other computer manufacturers came up with simulation to test which moving average period would best forecast the future. They went back in time, pretended that demand had not already occurred, and used different moving average periods to forecast the future. Based on these simulations they chose the moving average period that an inventory management system would use.

Computer simulation is the fundamental concept in focus forecasting. Without simulation focus forecasting would still be simple, but it wouldn't work. Focus forecasting needs today's superfast computers, but it is deeply rooted in the past. It needs the trends, cycles, seasons, and noise from the recipe approach. It couldn't work unless exponential smoothing had forced inventory managers to put demand data on the computer. And it certainly would never work without the innovation of computer simulation.

Some of the concepts in this book are very simple, and you probably already know of them. Some of the concepts in this book are very simple, but you probably never heard of them. I was lucky. My experiences have allowed me to see many different sides of the inventory management problem and to draw on concepts from many areas.

My first love was data processing and computers. In the 1960s I wrote a software compiler that was used in this country and Europe. I understand the nuts and bolts of programming support for computerized inventory management systems. I've been through the wrath of line managers who cannot understand why programs take so long to be written. As a line inventory manager I've been frustrated with the seeming lack of urgency some data processing people display in completing projects on time. But overall, the most fortunate experience of my career occurred while I was manager of systems planning for a major Fortune 500 company. I directed a staff of management consultants who later became independent consultants and authors. During this period I met people outside the company who are now the recognized leaders in inventory management theory and operations research. What I learned from them allowed me to formulate the simple system in this book.

Many of the experiences I had were not positive, but they all taught me about inventory management and the value of a simple system. I remember working with an inventory consultant on the new development of a retail inventory management system for a group of department stores in Cincinnati. Together in 1965 we developed a retail inventory replenishment system that will never be duplicated. It was a monument to complexity.

The inventory management consultant didn't like retail sales patterns; retail sales volume was just too small to exhibit predictable statistical behavior. The management consultant used to say, "All retail inventory control was stuck in the extreme tail of the probability curve." This man's enormous capacity for research led him to a system that dealt with forecasting rare events like floods, droughts, and fatigue failures in metals. The system was from a book titled *Statistics of Extremes* by E. J. Gumbel. The management consultant reasoned that since retail sales were rare events, it was an ideal system for retail inventory control. Together we applied statistics of extremes to retail inventory forecasting. I don't believe any of us except the inventory management consultant really understood the detailed mathematics of the system. The final report was so complex that each page of computer output contained only one item for each store. There were five Cincinnati stores and five hundred stockkeeping units in each store.

I can't really say whether the system worked or not. It required the management consultant's personal attention to keep it running from week to week. We replaced the system in short order with a system used by a giant retail chain. They made it work. The system was the replace sales system developed originally at J. C. Penney. That negative experience taught me that simple systems are the only systems that work.

In the appendix are some systems that do work. They don't fit into the natural scheme of the focus forecasting inventory management system, but they do fit all sorts of special cases. There's the replace sales system developed by J. C. Penney. There's profile forecasting that uses small samples of voter results to predict nationwide elections. There's the ranking system used by E. I. DuPont to forecast sales for new items. And there are economic models such as the games played at Carnegie Mellon to teach their students about business life. There's a matrix analyzer approach for gathering and summarizing market research data. All these systems use simple concepts and they work.

This is a simple book about logical things. If I tell you about something that is complicated, it's probably because I don't really understand it myself. This book is based on things I've learned from IBM, Honeywell, and NCR. It's based on things I've learned from Arthur Anderson, Booz, Allen & Hamilton, Weintraub Associates, Coopers & Lybrand. And it's based on things I've learned from some really outstanding inventory consultants, like Bob Brown, Ollie Wight, Dr. Michael Shegda, Dr. Joseph Bowman, Jerry Cantor, John Heinz, Bob White, Dave Seversky, Jack Sparrow, Eugene Baker, and many others. And from the business side and the people side, it's based on the experiences I've had working for general managers like Jack Field, John Moriarty, Phil Lamoureux, and John Berryman. I believe this book looks into the future of inventory management. And if I've been able to see a little further than some others, it's because I've had the opportunity to stand on the shoulders of these giants.

Focus
Forecasting

The Concept of Focus Forecasting

Inventory management is a tough job. It is based on forecasts of the future. The forecasts are always wrong, so the inventory management is never right.

I'm the inventory manager at American Hardware Supply. I've been here about five years. Now that may not sound like much of an accomplishment, but on the average an inventory manager at American Hardware lasts two years.

If you're like me you're probably thumbing through the first chapter of this book to see whether it's got something of value for you. It does. It's a book about a revolutionary new forecasting concept called focus forecasting and about the inventory management system that focus forecasting is a part of.

The concepts are brand new. Focus forecasting is not a new way of quadruple smoothing. It's not a new way of adjusting alpha factors. In fact it has nothing to do with exponential smoothing at all. It replaces exponential smoothing. Focus forecasting is a logical way to have the computer estimate the coming demand for an item. In my company focus forecasting forecasts 100,000 stockkeeping units every month. The items include hand tools, swimming pools, electric razors, fly swatters, fly rods, paint, and practically everything else you'll ever find in a hardware store. It's a proven system that's been part of an inventory management system since December 1972.

I believe the management of inventory is the management of a business. When you read in the *Wall Street Journal* that this company or that company went out of business, you'll usually find that it's because they had major inventory excesses. When you ask why customers buy from a particular supplier, service is always at the top of the list. In most companies inventory is from 33 percent to 50 percent of their total investment. Inventory turnover in large measure determines their return on investment. Inventory management is about the management of business. It should certainly interest inventory managers, but it should also interest anyone who is a student of business management.

If you are the general manager of some company, here are some of the secrets for getting off the roller coaster ride between low points in your customer service and excesses strangling your cash position. If you are the finance manager, here's a way to evaluate the inventory plans that come to you for cash resources. If you are in data processing, here is a simple computer forecasting system that works. If you are a student or a teacher, here are some new concepts for solving the age-old problem of inventory management.

This whole book is based on three principles: People will only use simple systems. Simple systems are the only systems that work. The only measure of any system is how well it works.

This chapter and the following chapters will tell you about a revolutionary forecasting concept that's the basis of an inventory management system that works. But the real magic of this book is how this forecasting and inventory management system allows people to make decisions. It allows them to understand the tools that the computer is using, and it holds them accountable for inventory level and customer service.

Now you won't have to dig through this whole book to find out how the focus forecasting system works. Here it is.

HOW FOCUS FORECASTING WORKS

Our computer has seven simple forecasting strategies in its vocabulary. They are forecasting strategies that people use every day. Strategies like whatever we sold during the last three months, that's probably what we'll sell during the next three months. Or, whatever we sold last year in the next three months is probably what we'll sell this year in the next three months. Or, whatever percentage we're running ahead or behind last year on this item in the past three months is probably the percentage ahead or behind last year we'll run in the next three months.

These are the simple forecasting strategies. The computer selects the one best strategy to forecast this item at this moment in time through the process of simulation. Simulation means the computer goes back in time three periods and pretends they didn't happen. The computer projects those three periods using each of the simple strategies. Whichever simple strategy best projects the three periods that have already happened is the one the computer uses to project the future.

Focus forecasting goes through this simulation every time it forecasts. It's a dynamic simulation. To pick the one strategy that worked best the computer divides the forecasted sales by the actual sales. The closer the answer is to 1, the better the forecasting strategy. This is the whole focus forecasting approach. There are no exotic formulas like exponential smoothing, or least squares regression, or cumulative line ratios. The computer is picking the one best strategy for this item for this moment in time. This means there isn't any file maintenance. Nobody tells the computer this is a seasonal item, or this is an A item, or a B item, or a C item. Focus forecasting itself picks the one best strategy that is most likely to work in the future.

People understand these simple forecasting approaches. Focus forecasting even tells the person looking at the forecast which simple strategy it chose. This approach using the simulative abilities of the computer produces extraordinary results.

RESULTS OF FOCUS FORECASTING

Focus forecasting has produced these extraordinary results. The first one is in the accuracy of the forecast. Every month we add up all the 100,000 forecasts that the computer makes in dollars. We add up all the actual demand dollars that occurred during the month. In stable years when we compare the forecasted dollars to the actual dollars over any three-month period, the difference is less than 1 percent. As a matter of fact, in April, May, and June 1976 the difference between forecasted dollars and actual dollars was 0.3 percent.

In 1973 during the Arab oil embargo demand for energy-related items skyrocketed. Yet focus forecasting's projected dollars differed from actual dollars of demand by less than 10 percent. Immediately following the oil embargo this country went through a recession. Even during that period of rapidly declining demand focus forecasting was still less than 10 percent wrong. The secret, of course, was focus forecasting's ability to adapt. Each time the demand pattern changed, focus forecasting switched to the strategy that best predicted what was really happening. Nobody told the system, "Hey, there's an energy crisis." Nobody told the system, "Boy, we're in a recession." Focus forecasting just quickly switched to strategies that best predicted the future for individual items.

The second result that focus forecasting produced was the buyers' increased trust in computer projections. In 1971 and 1972 our company used exponential smoothing with seasonality for forecasting the future. This forecasting system would suggest buys for the 100,000 items per month. The buyers reviewing these computer decisions changed more than 50 percent of the computer's decisions. With focus forecasting as part of an inventory management system, the buyers change less than 8 percent of the computer's buying decisions. When you compare all the buys suggested by the inventory management system to the actual purchases, the difference is less than 5 percent. The buyers trust the inventory management system. We respect their judgment. Buyers can change any computer decision they disagree with; yet they change less than 8 percent.

The third result of focus forecasting is probably the most significant. These results occur because focus forecasting is accurate and because the buyers trust the system. For the fiscal year ended June 30 focus forecasting, the inventory management system, and the buyers achieved the highest service level in the history of the company. The inventory at the end of the fiscal year was within 0.5 percent of the target inventory level.

WHERE DID FOCUS FORECASTING COME FROM?

Focus forecasting has produced some extraordinary results: extremely accurate forecasts, simple buying decisions that people understand, and exceptional turnover and service levels. Focus forecasting is a new approach because it uses simple systems that people understand and the simulative abilities of the computer. But where did focus forecasting come from?

Over the years I've had experience with a number of more sophisticated approaches dating from the early computers of the late 1950s and early 1960s. Each of these previous approaches had something good in it, and focus forecasting draws on these approaches. Here are some of them—what I found good in them, which I used to develop focus forecasting, and the things I found lacking in them.

The earliest approach to computerized forecasting in inventory management that I know of was developed in 1959. I like to think of it as a recipe approach. The recipe approach defines all demand in terms of trends, cycles, seasons, and noise. To forecast the future the computer reads from a "recipe": take two parts trend, one part seasonality, a five-year cycle, and sprinkle liberally with safety stock to cover noise.

Trends in forecasting are the consistent change in demand from one period to the next. For example, the trend is up two units a month, or the trend is up 1 percent a month. There are up trends, down trends, fashion trends, economic trends, just about every sort of trend. There really are trends in all demand. Trends are most pronounced in demand of high unit-volume items. In low unit-volume items trend is unpredictable, inconsistent, and unreliable. Focus forecasting uses the concept of trend only when it worked for an item in the past three periods. Figure 1.1 shows some graphic examples of trends from the *Wall Street Journal.*

Cycles are long-term curves that rise and fall over a period of years. Some say the cycles of winter goods like snowmobiles, ski wear, heaters, and fuel follow the cycles of sunspot activity. Every eleven years or so the sunspot activity reaches its low ebb and winters are very cold. During cold winters, winter goods sell very well. There are other cycles like the bicycle cycle. Every so many years bicycles become popular. After a few years their popularity dies. Nobody can predict when a bicycle cycle will occur, but when they do occur they should be recognized as cycles. Cycles affect

Figure 1.1 EXAMPLES OF TRENDS FROM THE *WALL STREET JOURNAL*

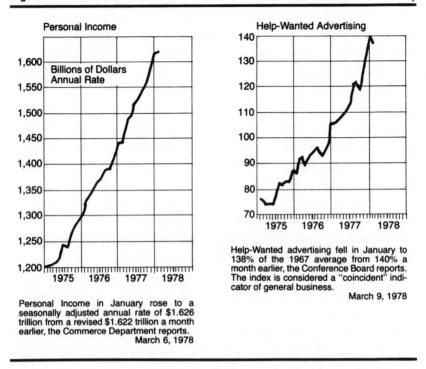

Personal Income

Billions of Dollars
Annual Rate

Personal Income in January rose to a seasonally adjusted annual rate of $1.626 trillion from a revised $1.622 trillion a month earlier, the Commerce Department reports.
March 6, 1978

Help-Wanted Advertising

Help-Wanted advertising fell in January to 138% of the 1967 average from 140% a month earlier, the Conference Board reports. The index is considered a "coincident" indicator of general business.
March 9, 1978

overall sales and the sales of broad categories of goods. Later you will see how the inventory management system of which focus forecasting is a part uses cycles in inventory planning.

Seasonality is a curve that repeats itself every year. The high point of the curve represents the peak sales period during the year. The more seasonal an item is, the greater the spread between the high and low points in the year. Usually seasonality is shown as the percentage of the total business that occurred during each period of the year. Christmas cards are obviously very seasonal. Grass seed demonstrates a double-humped seasonality, one for the spring and one for the fall. Some seasonality is explained by weather, some by holidays, and some by special events like Super Bowls, or graduations, or harvests.

The problem with the recipe approach is that it relies on people's judgment to project seasonality. Somebody gives the computer a seasonality curve to use for forecasting categories of items. Some of the items demonstrate the expected seasonality. Some of them don't. Again, it is inconsistent and unreliable. Focus forecasting recognizes seasonality in a different way. It uses a seasonal approach when it worked for an item in the past three periods.

In the recipe approach anything that can't be explained by trend, cycle, or season is classified as noise. Noise is the unexplained apparent random fluctuation in this period's demand. If you add up the last two digits in the telephone numbers on a page in your directory and divide by the number of people on the page, you get an average two-digit number. The recipe approach would use that average two-digit number to forecast the last two digits of the next telephone number. Some of the last two digits are higher than the average and some are lower. The telephone directory numbers' last two digits don't have any trend, cycle, or seasonality; all they have is noise. The recipe approach always expects trends and seasonality, so when it encounters demand that is purely random it forecasts some extraordinary bloopers. Those bloopers make people who use the recipe approach distrust computers and the people who invented the system.

The recipe approach produces usable results when sales volumes are large enough to display trends, cycles, and seasonality. But thousands of low-volume items in the distribution and manufacturing industries only display noise. Attempting to use a straight recipe approach on these items is like pounding a square peg into a round hole.

Practically all the forecasting innovations since the mid 1950s have been improved methods for estimating trends, cycles, seasonality, and noise. Focus forecasting uses trends, cycles, and seasonality. But focus forecasting is a brand new concept. It uses the computer as a simulator; it uses the computer to pick the best strategy for this item for this moment in time. It is always searching for and finding the round peg for the round hole.

EXPONENTIAL SMOOTHING

Sometimes it seems that computers make mistakes, but 0.9999 of these mistakes are programming or program software errors. In 1959 the computer itself made mistakes. I still have emotional scars from some of the mistakes computers made in those days. Sometimes one of the computer's tubes would burn out, or it would read a 7 as a 9 because the punch card went in crooked, or it would punch some extra holes in an answer card because the computer magnets got sticky. It didn't make many mistakes, but the more data it processed and the more calculations it performed, the greater were the chances of error.

The recipe approach used a large volume of demand data, and when those early computers processed those large volumes of data they made significant mistakes. There was a man who did something about it. Bob Brown used exponential smoothing. He knew the mathematics of inventory control and he knew the limitations of early computers. His innovations in using the computer to solve inventory control problems go far beyond exponential smoothing, but he will always be known for exponential smoothing. No matter what role Leonard Nemoy plays in the future, people will always say "There's Mr. Spock, the man with the pointy ears from *Star Trek.*" And when people think of Bob Brown their first thought will be, "That's Bob Brown, the man who first used exponential smoothing for inventory management." Yet exponential smoothing is only an abbreviated way to calculate a moving average. In those early computers storing a twelve-month history to calculate a moving average was expensive, inaccurate, and dangerous. So Bob Brown used exponential smoothing. I'm sure that any of you who are students of inventory management know of exponential smoothing. In its simplest form it is probably the most widely used forecasting device in the world today.

Focus forecasting doesn't use exponential smoothing to approximate moving averages. Why? Well, computers today don't make mistakes. They are nearly 100 percent accurate. Yet all over the world people are still using exponential smoothing. If you have a choice of adding up the last twelve months' demand and dividing by twelve to get a twelve-month moving average, or using exponential smoothing with an alpha factor of 0.154 to get an approximation of that twelve-month moving average, why not just use the simple moving average? It's a lot easier to explain to a user that the twelve-month moving average is the past twelve months' demand divided by twelve than to go into the whole agony of what an alpha factor is and what a beta factor is and why you multiply the current demand by alpha and the previous moving average by one minus alpha.

So focus forecasting doesn't use exponential smoothing. But without exponential smoothing there wouldn't be any focus forecasting. There probably wouldn't be any forecasting at all. This simple device—an approximation of a moving average—allowed companies to use the computers of the late 1950s to make usable inventory management decisions on thousands of items. It was the foundation for all computerized inventory management systems. It was because of exponential smoothing that tens of thousands of companies went through the discipline of gathering data and putting it on their computers. Exponential smoothing created the data base for all subsequent forecasting systems including focus forecasting. Focus forecasting does not need a data base created by exponential smoothing. All it needs is an eighteen-month demand history for an item. But tens of thousands of companies now using exponential smoothing can use the same data base for a focus forecasting system of their own.

The recipe approach created the fundamental concepts for all forecasting systems. Exponential smoothing provided the data base for all forecasting systems. In the early 1960s people combined the two techniques and used them for computer forecasting of future demand. They used the simple moving average from exponential smoothing coupled with trends, seasonality, and noise concepts from the recipe approach. There were many such systems, but the most widely used was IBM's IMPACT for manufacturing companies. Retail companies used IBM's IMPACT and NCR's REACT. Later developments that refined seasonality and made exponential smoothing moving averages more sensitive led to the development of IBM's COGS.

A MAJOR BREAKTHROUGH: COMPUTER SIMULATION

Forecasting systems that used exponential smoothing coupled with the recipe concepts were run on many of the early computers. Because the early computers used drums for memory storage, they were relatively slow for processing large volumes of data. Second-generation computers used core memory, which was hundreds of times faster than drum memory and allowed the next development in inventory management systems.

The exponential smoothing approaches could use many different moving average periods for forecasting the future, so the question was, What moving average period should be selected? What moving average period would best forecast future demand for my company? These questions led to the use of computer simulation in forecasting systems.

Computer simulation really sounds like a mouthful, but all it means is that the computer pretended that a period of the past had not happened. It would go back in time and forecast these periods that had already happened. Then it would compare the forecasts of past periods to the actual demand. It would show the results of different moving average periods in relation to their accuracy in forecasting those prior periods. Second-generation computers were fast enough to allow companies to run these simulations on thousands of items. Some companies ran these simulations on all their items, but most companies ran these simulations only once or twice a year to select moving average periods to forecast their future demand.

The concept was excellent. If we had to gamble on an uncertain future why not at least pick a forecasting strategy that worked in the past?

Computer simulation was a major forecasting breakthrough, but it had limitations. First, just about the only parameter that could be changed in the forecast was the period of moving averages. Some of the more sophisticated simulations incorporated trend sensitivity, but overall the parameters that could be changed in the simulation were still limited. Another drawback of early simulations was that it took a long time to run a simulation, so simulations were made infrequently.

Focus forecasting embraces simulation completely. It uses the simulation concept every time it forecasts and chooses from a series of simple forecasting strategies. The simulation approach allows focus forecasting to choose one simple strategy for this item at this moment in time.

Over the years computer output devices have become faster. Printers used to run 100 to 150 lines per minute; now they can print 1200 lines per minute. Machine readable computer output was produced at the rate of 6400 characters a second. It is now produced at the rate of 356,000 characters per second. The major increase in computer output speed, however, was in its central processing unit. Whereas it used to take thousandths of a second to perform calculations, it now takes billionths of a second. It's this tremendous surge in internal processing speed that makes focus forecasting possible.

Focus forecasting simulates every time it forecasts. In other words it forecasts our 100,000 stockkeeping units eight times every month to find the one forecasting strategy for each item at that moment in time that will work the best for the future. This dynamic simulation approach would have melted down the drum computers of the late 1950s. Today, however, it is the most efficient use of third-generation computers. It uses the strongest characteristic of third-generation computers—their internal processing speed.

Focus forecasting uses trend and seasonality concepts from the recipe approach. It can use the data base founded by exponential smoothing. It builds on the major innovation of computer simulation. This new approach is possible only now because of the speed of third-generation computers. It is deeply rooted in the inventions of the past. But because of its simulation approach, it is different from all statistical approaches of the past.

SIMULATION VERSUS STATISTICAL APPROACHES TO FORECASTING

Every computer forecasting system that I know of before focus forecasting depends on a statistical concept called normal distribution. The name of the game in these forecasting systems is to forecast the average value. Even if these systems forecast the average value, normal distribution produces as many errors on the high side as on the low side. These systems reject any forecast that is consistently high or low. Later versions of exponential smoothing called adaptive smoothing actually change moving average sensitivity if any consistent high or low bias is detected. Focus forecasting does not depend on normal distribution. Focus forecasting reacts to the real world and adapts to it.

Here is an example showing the difference between a statistical approach to forecasting and a focus forecasting approach. The example is flipping a coin. Don't skip over this because the conclusions from this traditional coin-flipping example are different from any you have ever read before. A schoolboy knows that the odds of flipping a coin are a 50 percent chance of heads and a 50 percent chance of tails. A statistical approach to forecasting heads or tails would be to call heads every time. Since there is a 50 percent chance of heads and a 50 percent chance of tails, that statistical forecast should be right at least 50 percent of the time. That statistical approach should satisfy the requirement for an unbiased forecast.

The problem is that in the real world normal distribution occurs only for heavy volumes of transactions or very long periods of time. Normal distributions do not occur with individual items or with individual people. Suppose a small boy actually flips that coin. He flips the coin and gets tails. Our formula loses. He flips the coin again and gets tails. Our formula loses. He flips the coin again and gets tails. Our formula loses. The statistical approach continues to forecast heads. After all there is still a 50–50 chance of heads. But with this particular small boy and with this particular coin, normal distribution may not apply. Maybe the small boy is not flipping the coin high enough. Maybe the coin itself has tails on both sides. Maybe the coin is bent and can't land tails down. If the small boy were my son I'd call tails. Focus forecasting would look at the recent past, and focus forecasting would call tails.

The laws of probability and normal distribution are tried and proven. They are right. Statistical optimization approaches to inventory management are wrong. Why? Normal distribution is based on large volumes of transactions and long periods of time. There is a rule in inventory control called the 80/20 rule. The 80/20 rule says that 80 percent of the volume in business is usually done by 20 percent of the items in a business. The other 80 percent of the items do not lend themselves to statistical forecasting techniques.

In inventory control we are not trying to have an excess for five years and a shortage for five years. In inventory control we are trying to determine the best stock level for this item at this moment in time. In inventory control we are not trying to have too much inventory for half the items and too little inventory for the other half of the items. We are trying to have the right inventory level for all items at this moment in time.

Now this is not just an idle theory. After repeated simulation many companies classify their items into A items, B items, and C items. A items are those 20 percent of the items accounting for 80 percent of the volume. B items have moderate volume. C items are a large percentage of the total number of items, but they account for a very small percentage of the total volume. Using statistical forecasting concepts for A items is fine. Using statistical forecasting concepts for B items sometimes works. For C items they rarely work. So most companies use computer statistical forecasts for their A items and human judgment for the others.

If you ask a company whether it has a computer forecasting system and the answer is yes, your next question should be what percentage of items that forecasting system controls. You will find that even though the computer may control 80 percent of the company's volume, it probably controls less than 20 percent of the items the company handles.

Focus forecasting does not depend on the statistical concept of normal distribution. Focus forecasting selects a strategy for a particular item at one moment in time. Focus forecasting makes the buying decisions for all 100,000 stockkeeping units in our company. Appendix 6 contains a more detailed discussion of statistical versus simulation approaches in making decisions about the future.

Later in this book we will discuss the difference between using statistical approaches and simulation approaches to setting safety stocks. But the main thing to remember here is that most forecasting systems in existence today use statistical approaches to forecast the future. These statistical approaches usually only work for high-volume A items. Focus forecasting uses simulation rather than statistical optimization approaches. Focus forecasting forecasts the future for all items.

TEMPTATIONS OF COMPLEXITY

There is a certain amount of status in not being understood. Almost every business discipline has a jargon that the in-people understand. For instance, inventory control has standard deviations, mean absolute deviations, double smoothing, sine and cosine seasonality waves, tracking signals, and time-phased reorder points. It's impressive to hear a group of people using these terms.

There's a real temptation to make systems complex to impress people. A new system that is fairly complex gives its author a certain amount of job security. Yet the people who must use systems will prefer a simple one to another which is difficult to understand. Ideally, everyone should remember that the first rule of system design is the KISS rule: Keep it simple, stupid.

It's a wonder that the KISS approach is so widely known and so little used in practice. In the preface of this book I told you about a retail inventory replenishment system that my staff developed for a group of department stores in Cincinnati, the monument to complexity. It certainly was impressive, but it failed miserably. It helped teach me that simple systems are the only systems that work.

Focus forecasting could use a series of exotic formulas to forecast the future. It could use exponential smoothing. It could use least squares regression. It could use cumulative line ratio forecasting. But it doesn't. It uses simple strategies that are easy to understand. Buyers looking at a forecast made by focus forecasting know how it works. They can look at a letter on the computer output and know which simple strategy focus forecasting chose. The buyers understand the forecast and the buying decision so the buyers control the decision. Because they understand the system and control the decision, they make the system work.

I hope you are not one of the people who admire complexity, who like mysteries. I am sure you have never seen an expensive consulting report without formulas, tables, indexes, spiral-bound papers, diagrams, and organization charts. Are they hard to read?

You bet. Some people think the more complex a system is, the better it must be. Don't believe it. If anybody takes more than an hour to explain something to you, either he does not understand it himself or he wants to impress you.

Don't measure the worth of a forecasting system by the complexity. The measure of a forecasting system is how well it works. Weather forecasts used to be summations of jet streams, cold fronts, high and low pressure areas, and sunspot activity. Now they take a picture of the world from a satellite. If there are clouds to the west it is probably going to rain. The satellite picture makes a point. If you see something clearly, you can explain it simply.

When I first explained focus forecasting to people I respect I was sheepish. I could explain the whole system in eight minutes. Use simple forecast systems. Let the computer use the one that works the best. Use the simulation power of the computer. I was worried that they would laugh. But they didn't.

My friend Dr. Joseph Bowman of Carnegie Mellon University said a forecasting system that adapts a series of formulas to item demand will outperform any single-formula forecasting system, provided there are a sufficient number of formulas available to choose from. Ollie Wight said a forecasting system that uses the simulative abilities of the computer to forecast the future is a major advance over any statistical optimization approach. John Berryman, the General Manager of American Hardware, said that the recent past is the best indicator of the future.

SUMMARY

In this chapter we talked about some of the inventions of the past. Focus forecasting draws on these inventions. It uses the concepts from the recipe approach, it can use the data base from exponential smoothing, and it uses the early innovation of simulation. It recognizes the limitations of statistical approaches to optimization and chooses instead simple simulation approaches. It avoids complexity wherever possible. It follows the basic rules: (1) People will only use simple systems. (2) Simple systems are the only systems that work. (3) The only measure of any system is how well it works.

As a summary of this chapter, here's a picture that tells the whole story of a forecasting system that people understand—a happy buyer!

(Actually this is Bob Koehler, Departmental Merchandise Manager at American Hardware Supply Company.)

Chapter 2

The Mechanics of Focus Forecasting

By now you must be saying to yourself, "Just how does that focus forecasting work?" This chapter covers the mechanics of a focus forecasting system. If you don't like details you may be thinking, "I'll skip over the nitty gritties." I hope you don't. I've made it as painless as possible for you. Go through the example in the next section. It's really the only way to understand how focus forecasting works. You should have some fun trying to outguess the focus forecasting example. People all over the country have played the "Can you outguess focus forecasting" game. Of course, it gets harder all the time. Focus forecasting has picked the brains of a lot of "winners" for the strategies it uses now.

This chapter also gives you some practical do's and don'ts for developing your own focus forecasting system. How do you forecast new items? Is it possible to forecast extremely seasonal items like Christmas tree bulbs, snow shovels, and fly swatters? How do you handle that weird demand that pops up every now and then in your item demand history? How do you include the wisdom and experience of your buyers and inventory planners in your forecast strategies? How do you get people to trust your forecast system? The section called "Some Practical Do's and Don'ts" answers these questions.

Brace yourself. Here are the mechanics of the system.

THE MECHANICS OF THE SYSTEM

The best way to illustrate the mechanics of focus forecasting is to use it to project the demand for an individual item. Look at the

demand in figure 2.1 for a broiler pan. You see there the demand
listed for an eighteen-month period, January through December of
last year and January through June of this year. Demand is the
number of units that our customers ordered. It is not necessarily the
number of units that we shipped. Before you go any further, pretend
you are a buyer or an inventory planner and project the demand for
July, August, and September.

Figure 2.1   DEMAND IN UNITS FOR A BROILER PAN

	Jan	Feb	Mar	Apr	May	June	July	Aug	Sept	Oct	Nov	Dec
Last year	6	212	378	129	163	96	167	159	201	153	76	30
This year	72	90	108	134	92	137						

   This demand in units is real. It's right from our company
computer records. The demand for July through December of this
year has already happened. Make sure you try to guess the demand
for July, August, and September. You won't be alone—all our
buyers have played this game. Inventory consultants all over the
United States have played this game, and over five hundred people
at the Atlanta APICS Conference played this game. The best way
for you to appreciate the simplicity and accuracy of focus forecasting is to try to outguess the focus forecasting system.
   Write down the approach you used to develop the forecast for
July, August, and September. If you beat or come close to the
accuracy of focus forecasting, I suggest you include it in your own
forecasting system.
   Now we will develop a focus forecasting system of our own, and
we'll use this system to make three forecasts. We'll forecast July,
August, and September at the end of June. We'll forecast August,
September, and October at the end of July. We'll forecast September, October, and November at the end of August. We'll be moving
forward in time as we forecast.
   It would be awkward to use the names of the months and the
"this year, last year" wording. So to follow the illustration more
easily, we'll develop some simple notation to talk about the periods.
We'll number the months 1 through 12. To distinguish between last

year and this year, we'll call last year LY and this year TY. To distinguish between forecasts and actual results we'll call the forecast F and the actual A.

Look back at the demand in units for a broiler pan. If we wanted to refer to the demand for January through March of last year, we would say A LY 1 through 3. That would mean actual demand, last year, months 1 through 3. Just to make sure you understand this simple notation, find the demand for A TY 4 through 6. If you got an answer of 363, you understand the simple notation. If you got any other answer, you'd better read back through the preceding paragraph one more time before you continue.

Now we have an item to forecast and we have notation to refer to it. I asked you to forecast July through September. That would be F TY 7 through 9. Let's develop a focus forecasting technique to forecast the next three months' demand.

First we have to come up with at least two alternate strategies. If you remember, the recipe approach explained all demand in terms of trends, cycles, seasons, and noise. It would be a good idea to pick one strategy that recognizes a trend and one strategy that recognizes seasonality.

Moving averages recognize trend to some degree. If we picked a strategy that said, "Whatever the demand was in the past three months will probably be the demand in the next three months," that would be a strategy that recognized trend but not seasonality. If there was an uptrend each time we moved forward one month, the past three-month moving average would increase. In this way it would recognize trend. If we used "Whatever the demand was in the past three months will probably be the demand in the next three months," this strategy would not recognize seasonality. Seasonality expects that demand will follow the same pattern year after year. "Whatever the demand was in the past three months will probably be the demand in the next three months" ignores last year's demand pattern. So our strategy A will be "Whatever the demand was in the past three months will probably be the demand in the next three months."

Now we should have a strategy that recognizes seasonality. Any strategy that relates a projection of the future to last year has an element of seasonality in it. A simple strategy would be "Whatever percentage increase or decrease we had over last year in the past three months will probably be the percentage increase or decrease over last year in the next three months." This strategy recognizes seasonality. So strategy B will be "Whatever percentage increase or decrease we had over last year in the past three months will probably be the percentage increase or decrease over last year in the next three months."

The strategies we use in our company have seasonal and non-seasonal characteristics. We have seven strategies. It would be difficult to illustrate focus forecasting using all seven strategies. So to simplify, we will show you the mechanics using just two strategies to project the demand for the broiler pan. Surprisingly, even these two strategies do a remarkable job of projecting broiler pan demand.

Strategy A is "Whatever the demand was in the past three months will probably be the demand in the next three months." In the notation this formula is A TY 4 through 6 = F TY 7 through 9.

Strategy B is "Whatever percentage increase or decrease we had over last year in the past three months will probably be the percentage increase or decrease over last year in the next three months." In the notation that is (A TY 4 through 6 divided by A LY 4 through 6) times A LY 7 through 9 = F TY 7 through 9.

So now we have two simple strategies for our focus forecasting approach. Let's see how the computer would determine which strategy to use to forecast demand for July through September. The computer pretends that the past three months have not already happened. Then it uses strategy A and strategy B to project those three months. It compares the accuracy of the two forecasting approaches to the actual demand for those three months. Which-ever strategy is most accurate is chosen to forecast the next three months. The computer picks the strategy that comes closest to forecasting the actual demand. If one strategy is off 10 percent and the other strategy is off 20 percent, it picks the strategy that is off 10 percent. Let's look at the arithmetic of this approach for strategies A and B.

Testing the Accuracy of Strategy A

"Whatever the demand was in the past three months will probably be the demand in the next three months."

A TY 1 through 3 = F TY 4 through 6
1 – (F TY 4 through 6 divided by A TY 4 through 6) = the percentage of accuracy of strategy A

In other words, if we use January, February, and March of this year to forecast April, May, and June of this year, what is the percentage difference?

Substituting in the notation we have

72 + 90 + 108 = 270
1 – [270 ÷ (134 + 92 + 137)] = 25.6%.

In other words strategy A produces a forecast that is short 25.6 percent.

Testing the Accuracy of Strategy B

"Whatever percentage increase or decrease we had over last year in the past three months will probably be the percentage increase or decrease over last year we will have in the next three months."

(A TY 1 through 3 divided by A LY 1 through 3) times A LY 4 through 6 = F TY 4 through 6
1 – (F TY 4 through 6 divided by A TY 4 through 6) = the percentage of accuracy of strategy B

In other words, if we use the percentage of increase or decrease of January through March of this year over last year and apply that percentage increase or decrease to April, May, and June of last year to forecast April, May, and June of this year, what is the percentage difference?

Substituting in the notation we have

[(72 + 90 + 108) ÷ (6 + 212 + 378)] × (129 + 163 + 96) = 176
1 – [176 ÷ (134 + 92 + 137)] = 51.5%.

In other words, strategy B produces a forecast that is short 51.5 percent.

Based on this simulation we see that strategy A comes closer than strategy B, so to project July, August, and September we will use strategy A: "Whatever the demand was in the past three months will probably be the demand in the next three months." Or, in the notation, A TY 4 through 6 = F TY 7 through 9. Substituting in the notation gives 134 + 92 + 137 = 363. Our demand forecast for July, August, and September is 363 broiler pans. Figure 2.2 gives the actual demand in units for the broiler pan for the whole two-year period.

Figure 2.2 DEMAND IN UNITS FOR A BROILER PAN

	Jan	Feb	Mar	Apr	May	June	July	Aug	Sept	Oct	Nov.	Dec
Last year	6	212	378	129	163	96	167	159	201	153	76	30
This year	72	90	108	134	92	137	120	151	86	113	97	40

The actual demand was 120 in July, 151 in August, and 86 in September, a total of 357. Strategy A forecasted demand of 363. It was high by only 6 units! At the end of July we would go through the same procedure again to select the strategy for forecasting August, September, and October. At the end of August we would go through the same procedure again to select the strategy for forecasting September, October, and November. Figure 2.3 gives the forecasts for June, July, and August.

You can see from the results in figure 2.3 that strategy A was chosen for all three forecast periods. If the broiler pan develops a seasonality pattern that more accurately projects the future than our strategy A, focus forecasting will switch to strategy B.

Figure 2.3 FOCUS FORECASTING

Forecast quarter	Strategy chosen	FTY 7–9 Forecasted demand	ATY 7–9 Actual demand
July–Sept	A	363 = 134+92+ 137	357 = 120+ 151 + 86
Aug–Oct	A	349 = 92 + 137+ 120	350 = 151 + 86+113
Sept–Nov	A	334 = ?	296 = 86 + 113+ 97

408 = 137+120+151

These are the mechanics of the focus forecasting approach. Appendix 7 uses exponential smoothing to forecast broiler pans for August, September, and October. Exponential smoothing forecasted 384 units for August, September, and October, 34 units high. Focus forecasting forecasted 349 units for August, September, and October, only 1 unit low. In all the simulations that we have run using different variations of exponential smoothing—single smoothing, double smoothing, adaptive smoothing—focus forecasting has significantly outperformed exponential smoothing.

With just two strategies, focus forecasting predicted broiler pan demand with remarkable accuracy. With seven strategies, focus forecasting accurately predicts demand for a hundred thousand different items.

Here are some rules for developing your own focus forecasting system.

SOME PRACTICAL DO'S AND DON'TS

Focus forecasting is a very flexible approach. You can see that in developing your own focus forecasting system you can choose from many strategies. Here we will review some of the practical do's and don'ts in developing a focus forecasting system. This section covers new-item strategies, strategies for extremely seasonal items, procedures for handling unusual demand patterns, and procedures for incorporating human judgment in all your forecasting strategies.

NEW-ITEM STRATEGIES

Focus forecasting does not need file maintenance. In other words, it doesn't need people to adjust seasonality indexes, to code A, B, or C items, to reset alpha factors, or to identify items as stable or seasonal, or to tell the computer that the item is new. When you are developing your own focus forecasting system, there is a real temptation to put in seasonal indexes for new items. There is a real temptation to tell the forecasting system what percentage of demand will occur in each month of the year for the new item. Don't do it! Let focus forecasting pick a strategy based on the recent past.

I think the best way to illustrate this is to look at a new item. Suppose that in May an insecticide that kills roaches is added. Sales for the first six months of the year are May 342, June 651, July 642, August 289, September 341, October 207. Focus forecasting doesn't know the insecticide kills roaches. It doesn't know that there are more roaches around in the summertime than in the winter. Focus forecasting doesn't even know it is summer. Focus forecasting does know sales are declining. It picks a strategy. Focus forecasting knows the item is new, the item had demand for only six months, so it prints an R (for review) that tells the buyer to look at this buy. This is the best way to handle new items in a focus forecasting system.

Let focus forecasting project new-item demand based on the recent past. Don't provide a seasonality curve for this new item. In other words, don't tell the computer that insecticides sell better in the summer than in the winter. This can cause more problems than it cures. In tha first year a new item will not follow the seasonality pattern of other similar items already in the line. The first year a new item is introduced, it obeys the law of pipeline filling. Sales start slowly as retailers gain awareness of the product. Then, in a sudden burst the new item gains in distribution as retailers build their initial inventory. Next, consumers become aware of the product. This cycle usually takes at least six months. Some heavily advertised and promoted items go through the cycle more quickly. Others go through the cycle more slowly. Overall the cycle looks a lot like a roller coaster, starting at the bottom, moving to the top. Figure 2.4 is a graph of insecticide pipeline filling. As retailers analyze the way the new item moves, finally, the normal pattern of inventory replenishment is experienced at the wholesale or manufacturer level.

Focus forecasting does not believe item seasonality until twelve months after the pipeline filling cycle. It believes the seasonality of an item after it is eighteen months old. Supplying a seasonality curve for a new item causes more problems than it cures. The law of pipeline filling distorts the new-item projections.

Don't put in seasonal indexes for new items. Do let focus forecasting point out to your buyer or inventory planner that this is

Figure 2.4   INSECTICIDE PIPELINE FILLING

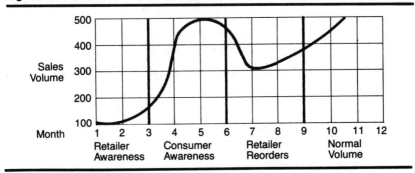

a new item and requires some review. Twelve months after the new item goes through its pipeline filling, focus forecasting will use seasonal approaches to project your sales. Until then you will be surprised at how well the recent past indicates the sales of new items regardless of seasonality.

EXTREMELY SEASONAL STRATEGIES

In applying focus forecasting to the 25,000 items in our line we came across the case of an extremely seasonal item—something like Christmas tree bulbs, snow shovels, or fly swatters. We affectionately call what we learned the "fly swatter philosophy." The "fly swatter philosophy" says that winter fly swatter demand has nothing to do with summer fly swatter demand. It is the exception to the rule that the recent past is the best indicator of the future. January, February, and March fly swatter demand has nothing to do with April, May, and June fly swatter demand.

The best way to illustrate the fly swatter philosophy is to look at a fly swatter demand pattern (fig. 2.5). Figure 2.5 shows that during January, February, and March this year fly swatter demand in-

Figure 2.5   FLY SWATTER DEMAND IN UNITS

	Jan	Feb	Mar	Apr	May	June	July	Aug	Sept	Oct	Nov	Dec
Last year	17	4	26	489	307	452	893	1617	204	10	3	42
This year	16	84	53	?	?	?						

creased 226 percent over last year. Does that mean that there will be a 226 percent increase in April, May, and June fly swatter sales? No! January, February, and March fly swatter demand has nothing to do with April, May, and June fly swatter demand.

The winter fly swatter demand is so small that increases and decreases are purely random. A statistical forecast, a recipe approach, or an exponential smoothing would look for a great fly swatter year. It would overforecast. Focus forecasting predicts that fly swatter demand will be the same as last year. Don't forget, there is no file maintenance in focus forecasting. Nobody told focus forecasting, "Hey, that's a fly swatter." There was no file maintenance. Focus forecasting knew that the fly swatter was a highly seasonal item. Figure 2.6 shows how it knew.

Figure 2.6 FLY SWATTER DEMAND IN UNITS

	Last year		Last year
January	17	April	489
February	4	May	307
March	26	June	452
	47		1248

Last year's January, February, and March fly swatter demand was less than 4 percent of last year's April, May, and June fly swatter demand. If last year's January, February, and March demand for wood screws was only 4 percent of last year's April, May, and June demand, focus forecasting would treat the wood screws just as though they were fly swatters. The demand determines the seasonality, not the item. The fly swatter philosophy works for lawn mowers, ice skates, Christmas tree bulbs, sleds, shotguns, swimming pool covers, and many other highly seasonal items. The beauty of the fly swatter philosophy is that it selects a forecasting strategy using simple logic. It selects a forecasting strategy by looking at the previous year's demand pattern. There is no file maintenance. There are no extraordinary bloopers in projected future demand. Don't forecast extraordinary bloopers on your extremely seasonal items. Do use the fly swatter philosophy.

HANDLING UNUSUAL DEMAND

Focus forecasting uses historic demand to project the future of an item. Sometimes the demand pattern looks unusual. Generally three things cause unusual demand patterns: inflated demand, outright order errors, and unexpected sales increases.

For whatever reason, if you run out of stock on an item, the demand on that item increases almost geometrically. In our company we do not backorder items. Our customers order once a week. In the course of a month they could reorder an item that was out of stock four times. If nothing was done to correct the demand generated by reordering, the demand would look extremely high. This extremely high demand would make extraordinary demand projections in the future. We call this inflated demand.

Some companies have tried to solve this problem by using sales rather than demand in their historic records. This creates the opposite problem. If you're using sales to project the future, as soon as you run out of stock it appears that sales suddenly stop and remain at zero until that stock is replenished. A forecasting system needs accurate demand numbers to forecast from.

Shortages can create inaccurate demand figures. A classic example of this was the energy crisis of 1973. In October 1973 locking gas caps sold at the rate of 80 per month. Then the Arabs shut off the oil supply. In October 1973 syphon hoses sold at the rate of 80 per month. Gas stations closed on Sundays. In November 1973 locking gas caps sold at the rate of 800 per month. Syphon hoses sold at the rate of 2400 per month. You can see in this example that demand on these items was ten to thirty times what it was before the energy crisis. The moral is that in times of shortage, demand is tremendously inflated. If we could have shipped 800 locking gas caps a day, demand for locking gas caps would have fallen off. If we could have shipped 2400 syphon hoses a day we probably would have shipped 2400 syphon hoses for only one day. The demand was inflated.

Here's the method we used for eliminating inflated demand. The first time a customer orders an item that we are out of, we count that order in demand. The second time the same customer orders the same quantity of the same item, we do not count that order in demand. If the customer orders a higher quantity than he had ordered originally, we count the additional quantity. If the customer repeatedly orders the same quantity for the same item, week after week, we do not count that demand until the item is back in stock or until some period of time has expired. In this way we are recording the demand that the customer would have ordered if the item had been in stock.

The best way to see how we eliminate inflated demand is to look at the following example:

October 24, 1973: Harry Saul orders 24 locking gas caps and gets them. So 24 locking gas caps are added to the October demand history.

November 1, 1973: Harry Saul orders 48 locking gas caps and doesn't get any. Then 48 locking gas caps are added to the November demand history.

November 8, 1973: Harry Saul orders 48 locking gas caps and doesn't get them. Nothing is added to November demand history.

November 15, 1973: Harry Saul orders 96 locking gas caps and doesn't get them. Then 48 more locking gas caps are added to November demand history.

November 22, 29, and the whole month of December 1973: Harry Saul keeps ordering 96 locking gas caps and doesn't get them. Nothing is added to the November demand history.

January 7, 1974: Harry Saul orders 96 locking gas caps and gets them. Nothing is added to January demand history.

January 14, 1974: Harry Saul orders 24 locking gas caps. Whether he gets them or not 24 locking gas caps are added to January demand history.

Figure 2.7 shows the demand history that results from the orders in this example. The logic here is to eliminate customer reorders for canceled merchandise.

There is a cost to setting up the system this way. The computer must keep a record of all canceled orders by customer and item. It must record the demand the first time an item is canceled. It must record any increase in an individual customer's reorder for an item even though the previous order was canceled. The computer must

Figure 2.7   HARRY SAUL'S HARDWARE / DEMAND HISTORY / LOCKING GAS CAPS

	October	November	December	January
Inflated	24	384	384	96
True	24	96	0	24

drop the item that was canceled the next time it is shipped to a customer or after a time has expired.

There is a cost for failing to set up the system this way. If the demand figures are wrong the forecasting system will be wrong, and not just this year. Next year it will expect the seasonality curve that doesn't exist. This problem is probably the most common flaw in computerized buying. If you don't fix it don't ask your buyers and inventory planners to rely on the computer forecasts. To fix this problem don't just use sales data. That's as bad as letting the inflated demand stay in the system. Do eliminate inflated demand by customer and by item.

The second and third types of unusual demand patterns are difficult to separate. The second type of unusual demand pattern is outright error. For example, a customer orders a wrong quantity of the right item, or the right quantity of the wrong item, or transposes numbers.

The third type of unusual demand pattern occurs when sales increase unexpectedly. This can happen for any number of reasons. In 1973 tornados raked the United States. People bought more pumps, water tanks, and fuses. In the same year consumers boycotted meat. Charcoal sales slumped. Vegetable prices sky-rocketed. Everybody grew a garden. One man even grew a garden on the top of a New York City penthouse. Sales of canning supplies and freezer bags were unreal. Remember the energy crises? Did you buy a gas can? A wood heater? An electric blanket? Everybody else did. And then what about the Christmas that was dark. Do you remember when President Nixon outlawed outdoor Christmas lights? Christmas light sales slumped. Every year and every month something unusual happens. Something always happens. How do you separate these normal unusual demand patterns from ordering errors?

Most statistical forecasting techniques use sales filters. They have formulas for eliminating unusual demand, but sometimes you don't want to eliminate unusual demand. These sales filters can't distinguish between the unusual demand you want to keep and the unusual demand you want to eliminate. Appendix 8 contains an example showing how statistical forecasts use sales filters to separate unusual demand. I don't recommend that you use these sales filters. They cause more problems than they cure.

The best thing to do with an unusual demand pattern is to point it out to the buyer or inventory planner. A simple way of identifying unusual demand patterns doesn't involve square roots or standard deviations. A simple way is to look at the past three months' demand in relation to the prior three months' demand and demand in the same three months of last year. If demand in the last three months is more than 200 percent of the prior three months' demand *and* more than 200 percent of the same three months' demand last year, this is certainly an unusual demand pattern.

If the last three months' demand was more than 200 percent of the prior three months' demand, this item has had an extraordinary increase in demand. If, however, in the same three months last year it had the same extraordinary increase in demand, it could just be a seasonal item. But if an item had an extraordinary increase in demand over the prior three months and an extraordinary increase in demand over the same three months last year, then someone has probably ordered the wrong quantity of an item or the right quantity of the wrong item.

Focus forecasting points out items with unusual demand patterns for the buyer or inventory planner's review. This is the simplest way to handle unusual demand patterns. It respects people's judgment and their role in the forecasting system. Because focus forecasting respects people's judgment, the people respect the tool they are using. Don't filter out unusual demand. Do point out unusual demand to your buyer or inventory planner.

USING PEOPLE TO DEVELOP STRATEGIES

The best way to develop a focus forecasting system is to let people participate in the development of the strategies. We do this by playing the "Can You Outguess Focus Forecasting" game with all our buyers. We simulate 2000 of our items and go back in time two years. We let focus forecasting make forecasts for every month in the past six months. We sit down with the individual buyers and cover up the forecasts and actuals for the six months that have already happened. Both the buyer and focus forecasting project the demand for the coming period. This is the major source of new input and changes to the focus forecasting system. If the buyer's strategy works better than focus forecasting on a number of items, we actually incorporate it into the focus forecasting strategy vocabulary. We resimulate all 2000 items. If the net performance with the added strategy is an overall improvement, we keep that strategy as part of the focus forecasting system. In this way people who use the system help create the system. Because they help create the system they understand the system and they trust it.

We play this game with the buyers every three months. We play this game with all new people in the buying department. Even the personnel manager in our company has played this game. When you develop a focus forecasting system of your own, don't play the game with operations research people and mathematicians. Play the game with the people who are going to use the system.

About 2000 people outside our company have played "Can You Outguess Focus Forecasting," including Ollie Wight, Walt Goddard, Dr. Michael Shegda, and Dr. Joseph Bowman. Some beat the system, some didn't. I'm not going to tell you who won and who didn't. But I think you can see that playing the game gets people involved.

Figure 2.8 gives examples of two people, Vaughn Hayes, Purchasing Coordinator at Lowe's Companies, Inc., and Dick Cantley, Purchasing, Church's Lumber Yards, who played "Can You Outguess Focus Forecasting." One person outguessed the system, the other didn't. But both of them now understand the system and respect its forecasts.

DEVELOPING TRUST IN THE SYSTEM

There are two rules for any strategy you add to your focus forecasting system. The number one rule is that it must be simple.

Figure 2.8 WHO OUTGUESSED FOCUS FORECASTING?

Power Tool Accessories

	Dick Cantley	Focus forecasting	Actual demand
December	40	40	42
January	40	31*	27
February	40	18*	24
March	36	35*	34
April	37*	34	41
May	20*	19	39
Total	213*	177	207

Oak Alkaline Floor Paint

	Vaughn Hayes	Focus forecasting	Actual demand
August	100	76*	33
September	45	40*	34
October	40	35*	27
November	23	20*	6
December	15*	14	33
January	20*	15	22
Total	243	200*	155

*The closer guess

People will be looking at a strategy used on thousands of items in your company. It must be readily understandable to them. The second rule is make sure that the strategy that was chosen is obvious to the person using the forecasting system. Make sure that the strategy is transparent. In our focus forecasting system, each one of our seven strategies has a letter A through G. This letter prints right on the buying report. The buyers are well aware of what each letter stands for. They can see what strategy was chosen and how it works.

When things are going well in inventory control it really doesn't matter what system you use. Everybody trusts it. The real test is when things are not going so well, when you are in an energy crisis or a recession. It's during these periods that people using the forecasting system must understand the computer's logic.

Simplicity and transparency are the keys to maintaining the trust of the people who use the system during periods of uncertainty. Do not add exotic formulas like exponential smoothing or least squares regression or cumulative line ratio. Keep the strategies simple so that they will be remembered and understood every day. Do not develop your focus forecasting system somewhere in the back room as a great mystery. Share its development with the people in your company. Make sure the system is transparent. It must always tell the people who use it what it's doing and why.

SUMMARY

Focus forecasting uses simple strategies. Some of those simple strategies can be suggested by forecasting techniques of the past like the recipe approach. You have seen focus forecasting in action on a broiler pan. The major development of a focus forecasting system for your own company should come from the people who use the system. Focus forecasting does not require file maintenance. Focus forecasting even produces usable results on new items and handles extremely seasonal items like fly swatters, Christmas tree bulbs, and snow shovels.

You've seen a procedure for eliminating inflated demand. Inflated demand is one of the major flaws in many forecasting systems. Inflated demand distorts demand history. In any demand history there are some unusual sales patterns. Focus forecasting relies on people's judgment to determine which unusual demand patterns are due to simple errors and which are just due to current economic conditions.

As part of an inventory management system focus forecasting not only makes individual buying decisions for every item in the company's line, it also provides information for total inventory planning and for expediting needed merchandise. Focus forecasting is primarily involved in finished goods inventory management.

In the next chapter we will review the six basic steps in the focus forecasting inventory management system. In our company we call finished goods inventory management "How to Keep Stock."

Chapter 3

The Finished Goods Inventory Management System

Finished goods inventory management involves most of the people in a company. There is a lot to it. This chapter tells you about the people who use the finished goods inventory management system. You have to understand their role in the company before you can understand their part in using the system. This chapter tells you what these people are trying to do with the system. It tells you the objectives of the finished goods inventory management system.

There are many new ideas in this finished goods inventory management system. This chapter will give you the six basic steps in the system and will help you see how each of these new ideas improves customer service and inventory turnover. This chapter is an overview of the entire finished goods inventory management system. It should provide a framework for your understanding of the following chapters.

WHAT IS FINISHED GOODS INVENTORY MANAGEMENT?

Finished goods inventory management is the whole inventory management job in wholesaling companies and in retailing companies. In manufacturing companies finished goods inventory management is at least one-third of the inventory management job. Yet I've seen little recent material on finished goods inventory management. Most of our current writing is about material requirements planning. Material requirements planning has helped thousands of manufacturing companies do a better job of meeting unpredictable customer demand, but it isn't the whole answer.

Material requirements planning is a scheduling system that matches the actual customer demand with the finished goods inventory. It calculates the planned requirements and converts them into production and raw material requirements. People who use material requirements planning properly have a powerful tool for reacting to change and to errors in their original forecasts and inventory plans.

It's great to be able to react to changes in plans. But it's also great to have finished goods inventory plans that don't require as many changes.

Finished goods inventory management includes planning the total inventory size and the total customer service level. It is the strategy for setting safety stocks to get the most customer service mileage from a given dollar invested in inventory. Finished goods inventory management determines what orders are needed first and helps insure the delivery of needed goods. It's the whole process of measuring the overall inventory management performance.

Finished goods inventory management doesn't replace material requirements planning; it works side by side with material requirements planning. In a manufacturing company it provides the input into the initial material requirements plan. In a wholesaling company it generates plans of future purchases that suppliers can put into their material requirements plan. Finished goods inventory management is a better way to plan for the future. Material requirements planning is a better way to meet the changes required in those plans. Side by side, finished goods inventory management and material requirements planning can produce extraordinary inventory management results. Over the last ten years our educators and our consultants have spent a great deal of time spreading the use of material requirements planning. For example, figure 3.1 shows an illustration of material requirements planning from Ollie Wight's recent book.

Figure 3.1   MATERIAL REQUIREMENTS PLANNING (from page 34 of *Production and Inventory Management in the Computer Age,* by Oliver Wight)

		Weeks							
Master Schedule		1	2	3	4	5	6	7	8
Bicycles		40	0	50	0	0	60	0	60

Lead Time = 4, Order Quantity = 100

MRP — Handlebars

Projected Requirements		40	0	50	0	0	60	0	60
Scheduled Receipts				120					
On Hand	60	20	20	90	90	90	30	30	-30
Planned Order Release					100				

USERS

Finished goods inventory management is the entire inventory management job in wholesaling companies. I'm going to describe the finished goods inventory management system in a wholesaling company because I am the inventory manager in a wholesaling company. You should be able to apply these same concepts to retail or manufacturing finished goods inventory management.

Figure 3.2 shows the people who use the finished goods inventory management system. You should be able to match up their duties and responsibilities to people with like jobs in retailing and manufacturing companies.

Figure 3.2   THE PEOPLE WHO USE THE FINISHED GOODS INVENTORY MANAGEMENT SYSTEM IN A WHOLESALING COMPANY

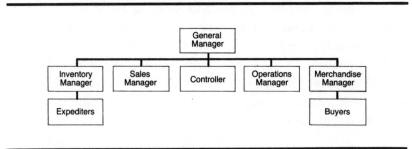

Inventory Manager

The inventory manager is responsible for three things:

1. Keeping a satisfactory customer service level; in other words having the goods available in inventory when the customers want them.
2. Keeping down the size of the overall inventory. Inventory eats up cash and inventory eats up space.
3. Buying goods in such a way that the cost of ordering goods is balanced against the cost of storing goods. The inventory manager must try to reduce the workload on the company's warehouses and get maximum quantity discounts and minimum freight costs.

The inventory manager determines inventory goals for the budget period. He decides how much to buy in total dollars. He identifies inventory excesses for disposal. He creates and controls the finished goods inventory management system. He wants realistic sales budgets. He wants fast-moving high-turnover items in his inventory. He relies on the other departments in the company to do their part in keeping customer service high and inventory levels low. He measures his own performance and the performance of each department in the company as it relates to customer service and inventory investment.

Sales Manager

The sales manager is responsible for increasing sales. He's responsible for adding new customers and for selling more goods to existing customers. He's responsible for making the initial forecast of how much will be sold in the budget period and for making those sales budgets happen.

When the sales manager looks at an inventory plan his biggest concern is that the inventory manager will have enough inventory to maintain a high customer service level. The sales manager is the person who says we can't sell out of an empty wagon. He's the one who says let's err on the high side.

Controller

The controller is responsible for pulling together the plans that will produce the desired net profit for the company. He challenges the sales manager's sales forecast as too optimistic or too pessimistic. He matches the company's operating expense budgets with the sales plan. He measures the operating expense performance of every department against their budget. In the finished goods inventory management system we'll see that in his financial statements the controller also measures the customer service and inventory size against budget.

The controller generally wants lower inventories and higher customer service. He wants better return on investment.

Operations Manager

The operations manager is in charge of the company's warehouses. He has the largest labor force under his control. He's responsible for effectively moving the goods from the vendor to the customer at the least possible cost. The operations manager wants a minimum level of inventory taking up space in his warehouses. He wants a minimum number of orders coming into his warehouses. Each order that comes into his warehouse requires labor to receive and to put away. The operations manager wants a high customer service level. He doesn't want his people wasting their time going past empty bins.

The operations manager can affect the overall customer service level tremendously. He must put away the needed goods when they arrive at the warehouse as quickly as possible. Sometimes his goal of keeping down operating costs conflicts with his goal of putting away the needed merchandise.

Buyers

The buyers work for the merchandise manager. They are responsible for picking the items in the company's line and for buying them at the least cost. They are responsible for setting prices. They are responsible for selecting vendors based on price, quality, delivery, and salability of merchandise. Because they determine how much of each item to buy, in large measure they are accountable for the customer service and for the size of the inventory in their department. When the inventory manager identifies excess inventories the buyers must determine how to dispose of them.

The buyers want simple buying strategies that work.

Expediters

The expediters work for the inventory manager, who holds them responsible for insuring the delivery of needed goods. As long as the goods are ready on time, the expediters don't really care when an order was written. Maintaining prompt customer service is the basis of an expediter's reputation, since the inventory manager monitors his ability to get delivery of needed merchandise.

General Manager

The general manager is responsible for customer service, increasing sales, net profit, return on investment, and low-cost operation. He depends on the department managers. He wants more of everything, overall improvement. He wants a finished goods inventory management system that supports the goals of all the managers.

These are the people who use this finished goods inventory management system in a wholesaling company. Their objectives are the same as their counterparts in retailing and in manufacturing.

OBJECTIVES

The people who use the finished goods inventory management system are trying to accomplish three objectives:

1. *Improve customer service.* Customer service means different things in different companies. It can mean the number of dollars the company ships at once on a customer's order, or the time it takes to ship a customer's order, or the number of dollars the company owes the customer. The customer service objective in this finished goods inventory management system is to ship as many items as possible at once on the customer's

original order. For example, a customer service objective might be to ship at once nine of the ten items that the customer ordered.

2. *Reduce the size of the inventory.* Companies measure the size of the inventory in different ways. Some companies compare this year's total inventory in dollars to last year's. Others compare the number of dollars in inventory to the number of dollars planned. Some companies measure inventory turnover, the total annual sales divided by the average inventory. The inventory size objective in this finished goods inventory management system is to reduce the months of supply needed to support the customer service objective. For example, an objective might be to get down to two months' supply at the end of June while continuing to ship at once nine of the ten items that the customer ordered.

3. *Reduce the total cost of ordering inventory and carrying inventory.* In manufacturing companies the cost of ordering inventory includes the cost of reviewing items for issuing orders, the cost of tracking orders through production, the cost of setting up and tearing down machines to produce items, and the cost of receiving goods in the warehouse and putting them away for sale. The ordering cost in a wholesaling company is similar. It includes the cost of reviewing items for buying decisions, the cost of keeping track of open purchase orders with vendors, the cost of receiving and putting away goods at the warehouses, and the cost of handling vendor invoices and mailing checks.

Inventory carrying costs are the same in both wholesaling and manufacturing companies: the cost of space, the cost of capital tied up in the inventory investment, the cost of taxes on the inventory investment, the cost of insurance on the inventory investment, and the cost of the risk of obsolescence of any item carried in the inventory.

Ordering costs and carrying costs are not linear. In other words, if we order one more item it doesn't cost an exact dollar amount to order that item. If we carry another hundred units of an item it doesn't cost an exact dollar amount to store that item. As a matter of fact, it doesn't cost anything to order or carry an item if we have the right mix of idle people, equipment, space, and money. But it can cost a great deal if we add orders to an overburdened production facility or if we add quantities to an overcrowded warehouse. The cost of ordering and carrying inventory depends on the level of capacity we are now using.

For these reasons the objectives for ordering costs and carrying costs in this finished goods inventory management system are expressed differently. The objectives are simply to reduce both the size of the inventory and the number of orders.

SIX BASIC STEPS

The first step in any planning process is to forecast sales. Usually it is the sales manager who makes the initial sales forecast. When he does so he's saying that he will find the necessary customers and sell enough to meet that forecast. Most sales managers are optimistic, and with good reason. Who wants a sales manager who forecasts no increase, or even a decrease, in sales?

The controller knows that the sales manager's forecasts are usually optimistic, so he challenges the sales manager's forecast until he is satisfied that it is realistic. Then the controller includes this sales forecast as the starting point of the budget for the coming period.

The inventory manager uses the controller's sales budget in his inventory planning. In large measure this sales budget determines the expense plans for all the departments in the company. It's sound business for the inventory manager to plan inventory levels based on the controller's sales budget. The inventory manager could forecast total sales since he knows many forecasting devices. But it's important to use the controller's sales budget because the controller is more likely to respect an inventory plan that's based on the company's sales budget.

So the first step in the finished goods inventory management system is: *The controller budgets the total sales.*

The inventory manager must determine the customer service

goal for each month of the budget period. To set this goal he has to answer some questions. How healthy is the inventory now? What was the customer service level last year? Is the coming budget period going into season or coming out of season? What are the company's objectives for customer service? These and many other questions must be answered to set the customer service goal for each month.

Once the inventory manager has determined the customer service goals for each month he can look at the company's historic performance month by month. Since it takes a certain number of months of supply to generate a customer service level, the lower the inventory relative to future sales, the lower the customer service level will be. The inventory manager can review the month-by-month relationship between months of supply and customer service achieved for two, three, four, and even five years.

This analysis helps the inventory manager plan months of supply for each month of the budget period. Once the months of supply have been determined it's relatively simple to calculate what the inventory level should be at the end of each month. If the inventory manager wants two months' supply at the end of June he simply adds forecasted sales for July and August. The total should be the total dollars in inventory at the end of June. This is the second step in the finished goods inventory management system: *The inventory manager plans total inventory levels.*

The inventory manager knows the sales budget, that's the outflow from inventory. The inventory manager knows the inventory levels by month. So now the inventory manager must plan the flow into inventory by month. The inventory manager adds the month's projected sales to the planned inventory balance and subtracts the inventory balance at the beginning of the month. This tells him how much to purchase to replenish inventory. He gives a schedule of these planned purchases to the operations manager who uses it along with the sales budget to plan labor requirements. So the controller has budgeted the total sales and the inventory manager has planned the total inventory levels. These first two steps are the planning part of the finished goods inventory management system.

Now, at the beginning of the new budget period the inventory manager must somehow purchase the total that he has budgeted. This finished goods inventory management system is the only one I know of that *automatically* marries financial planning to individual item buying. At the beginning of the month the finished goods inventory management system reviews the current supply of every item in the company's line and focus forecasting's projection of demand. It calculates the expected purchases for each item, extends all the purchases at their dollar cost, and totals them. It tells the inventory manager how many total dollars will be purchased at the current safety stock level. More than that, the finished goods inventory management system tells the inventory manager how to reduce the safety stock setting to lower the total dollar purchases or how to increase the safety stock setting to increase the total dollar purchases.

The inventory manager knows what the total dollar purchases must be and sets the safety stocks accordingly. This setting affects the purchase of every item in the company's line during the month. This is the third step in the finished goods inventory management system: *The inventory manager sets the total safety stock levels.*

At the appointed time during the month, each item is reviewed for a buying decision. The computer prints a report with a suggested buy based on the inventory manager's total safety stock setting. The buyers review each suggested buy and make any changes they wish to increase or decrease the suggested purchases. In this finished goods inventory management system the buyers change less than 8 percent of the computer's suggested purchases. The net result of all their changes is less than 5 percent. So by the end of the month the total purchases planned are very close to the actual purchases. This is the fourth step in the finished goods inventory management system: *The buyers buy the individual items.*

It's one thing to write purchases for the individual items. It's another thing to get delivery of the needed goods. The finished goods inventory management system tells the expediters the schedule of delivery. It tells them which orders are needed first. The expediters are responsible not only for getting the purchases from the vendor to the warehouse, but also for making sure that the goods move from the warehouse dock to the shelf for sale to the customer. The expediters' work is done primarily over the telephone. Expediters must have a personal rapport with the vendor's customer service representative to get their orders put on top of the pile, to get their goods shipped first. This is the fifth step in the finished goods inventory management system: *The expediters insure the delivery of needed goods.*

During the budget period the controller's financial statements show the actual and budgeted inventory levels at the end of each month as well as the actual and budgeted customer service levels. The inventory manager comments briefly on any variance from budget, but the inventory manager needs to evaluate performance in more detail. He measures each buyer's department not only on their inventory investment and customer service, but also on the amount of excess inventory they have accumulated. The inventory manager measures the sales budget against actual. He measures the planned purchases against actual. He measures the operations manager's performance in putting away the needed goods. He measures the expediters' ability to insure the delivery of needed goods. He measures each day, each week, and at the end of each month. He changes plans based on a comparison of actual and projected demand and a comparison of actual and expected delivery from vendors. The inventory manager's measurement of the total inventory management system provides the input for planning each month. This is the sixth step in the finished goods inventory management system: *The inventory manager measures performance.* Figure 3.3 is a flow chart showing the six basic steps in the finished goods inventory management system. Chapters 4 to 9 discuss these six basic steps. The rest of this chapter discusses some of the new ideas in those chapters.

CUSTOMER SERVICE AND INVENTORY SIZE

Many companies don't plan the total size of their inventory. Instead, they plan their customer service levels. They have ways of setting customer service objectives for A items, their best sellers; B items, their more moderate sellers; and C items, their worst sellers.

Figure 3.3   THE FINISHED GOODS INVENTORY MANAGEMENT SYSTEM

	How to Keep Stock	
Planning	1. *Controller* Budget total sales	Historical sales records
	2. *Inventory manager* Plan total inventory levels	Historical turnover and service records
Execution	3. *Inventory manager* Set total safety stock levels	Inventory position report
	4. *Buyers* Buy individual items	Buyers reports
	5. *Expediters* Insure delivery of needed goods	Expediting report
Feedback	6. *Inventory manager* Measure performance	Financial statements

The problem is that when these companies set customer service objectives, they affect the size of their inventory. Their finished goods inventory management systems are building safety stocks to meet their customer service objectives, but their inventory planning is not tied to their financial planning.

The finished goods inventory management system that we are describing works just the other way around. It begins with financial planning; it starts with the sales budget. The sales budget and the customer service objectives determine the size of the inventory, and the size of the inventory determines the size of the safety stocks. This gets the company off the roller coaster. In many companies the inventory manager increases inventory until the customer service level improves. Then the controller applies pressure because cash resources are strained. Then the inventory manager reduces the size of the inventory. The customer service level goes down. The customers start to scream, the general manager puts pressure on the inventory manager to build inventory. And the cycle goes on and on.

In this finished goods inventory management system the size of the inventory and the customer service level are planned. The safety stock is tied to the size of the inventory. The buying systems buy quantities of the individual items that will meet the financial budget for inventory level. It takes a lot of the emotion out of inventory management.

SAFETY STOCKS

The traditional approach to setting safety stocks is based on the laws of probability. The more forecast error, the higher the safety stocks. But most companies really want to remain in stock on their best-selling items. Most forecast errors occur in the poorer-selling items, so safety stock strategies based on the laws of probability put most of the safety stock in these poorer-selling items. Companies try to change this by classifying items A, B, and C. They try to put most of their safety stock in the A items, but they defeat their purpose by having their safety stocks based on the laws of probability. Laws of probability always put most of the safety stock in the worst-selling items.

Our inventory management system puts most of the safety stock in the best-selling items in the company. When the inventory manager sets total safety stock, he does not change this distribution of safety stock. Most of the safety stock remains in the best-selling items.

BUYER TRUST

Here are two ways companies try to get buyers to use finished goods inventory management systems.

Company 1. The company believes the system should not be changed. They instruct the buyers not to change the suggested buys. In these companies the buyers feel no responsibility for the size of the inventory or the customer service.

Company 2. The company allows the buyers to change the suggested buys. In many of these companies the inventory management system is not doing anything. The buyers are changing more than half the buying decisions. These companies really don't have an inventory management system; all they have is an information system that gives the buyers the information they need to make a buying decision.

In our finished goods inventory management system, the buyers can change any of the buy quantities suggested to them. Yet changes are rarely made, since the suggestions made by the system are based on the buyers' original logic. The buyers are responsible for the size of the inventory and for customer service, but they are helped by a powerful tool that makes accurate buys for them.

ORDER QUANTITIES

Many companies use economic order quantities. Economic order quantities supposedly balance the cost of ordering inventory against the cost of carrying inventory. The traditional economic order quantity formula requires estimates of inventory ordering costs and inventory carrying costs. These estimates are almost always inaccurate. Worse yet, inventory ordering costs and inventory carrying costs aren't linear. So these companies buy certain quantities at certain times based on information that is inaccurate and based on a formula that does not follow the actual cost structure in the company.

We will see a new way of balancing ordering costs and inventory carrying costs that does not require estimates of inventory ordering costs or inventory carrying costs. It reduces inventory ordering costs and inventory carrying costs by reducing the number of orders that are written and by reducing the total size of the inventory.

COMMUNICATIONS

Most companies just launch an order and expect delivery. When their inventories get too high they cancel orders. When their inventories get too low they launch enormous orders. In this finished goods inventory management system you will see a concept called distribution requirements planning. Distribution requirements planning carries the concept of material requirements planning into the customer–supplier relationship. This finished goods inventory management system provides a schedule of planned future purchases to suppliers, who can then use this information to plan their future production and raw material requirements. This is a powerful device for insuring delivery.

EXPEDITING

I've read a lot of books on inventory management, but I've seen very little on expediting. People view expediting as calling suppliers and getting information. That kind of expediting is a waste of time. Getting rid of such expediters would save a company money. Expediting means developing a personal rapport with suppliers. It's this personal rapport that separates the professional expediting staff from a staff of clerks.

MEASURING PERFORMANCE

Many companies measure performance. In finished goods inventory management it's necessary to measure at least three areas of performance simultaneously: inventory turnover, customer service, and gross profit. If the inventory manager measures just turnover the buyer will reduce inventory and reduce customer service. If the inventory manager measures just customer service the buyer will increase the inventory indefinitely. If the inventory manager measures turnover and customer service the buyer will sacrifice gross profit by reducing prices to increase sales, by discounting excess inventories prematurely, and by not taking the risk of adding new items.

We will see how the inventory manager in this finished goods inventory management system evaluates the buyer on inventory turnover, customer service, and gross profit.

SYSTEMS DON'T, PEOPLE DO

Most of the new ideas in this inventory management strategy are new systems concepts. Like most systems-oriented people I sometimes get carried away with the system itself. The systems in this book do literally make decisions. But they are just following the logic that some person programmed into the system. The inventory plans are worthless unless some person sets customer service goals. The computer-suggested purchases are dangerous unless some person reviews and changes those 8 percent of the items for which the computer is wrong. The expediting system can select the orders we need now. But some person must use personal influence to persuade a vendor to ship his company's goods first.

These new systems are powerful tools. Still, the secret to meeting inventory objectives is people. Systems don't, people do.

SUMMARY

People who use material requirements planning have an excellent tool for reacting to change. This finished goods inventory management system reduces the number of changes they are forced to make. People use this system to accomplish three objectives: (1) improve customer service, (2) reduce the size of inventory, and (3) reduce the total cost of ordering and carrying inventory.

There are six basic steps in this finished goods inventory management system:

1. The controller budgets total sales.
2. The inventory manager plans total inventory levels.
3. The inventory manager sets total safety stock levels.
4. The buyers buy the individual items.
5. The expediters insure the delivery of needed goods.
6. The inventory manager measures performance.

In our company we call these six steps "How to Keep Stock." In the chapters that follow we will discuss these six steps in more detail.

Chapter 4

How to Plan
Total Inventory Levels

When I first started to work as an inventory manager, I went to my boss for some guidance. "Well," he said, "you're the expert, but let me mention that over the years we've had some real problems with inventory. When the inventory gets too high you'll have our controller chasing you up and down the hall."

"I see," I said.

"Yes, in a growing company like ours, cash is a real problem," he said, "and worse yet, a heavy inventory puts a tremendous burden on the warehouse people. I don't want them in here complaining about too much of this and too much of that."

"Hm, hmm," I said.

"The general manager gets very upset when we start having cash problems. He'll push us to get rid of excesses. He'll really get tough when we have to dump merchandise at a loss."

"I understand," I said, and I got up to leave.

"There's one more thing," he said. "If the inventory gets too low you get fired."

THE INVENTORY PLANNING DILEMMA

Ask a controller what a satisfactory level is and he'll tell you lower than last year's inventory level. After a lengthy discussion on the reasons for needing more inventory dollars—the current shortage of raw materials, the cost saving of an economic purchase order quantity, the principles of hedging in a price-freeze environment, the optimism of the sales budget, the need for customer goodwill for long-term growth, a mystical reference to the eleven-year sunspot activity cycle—the controller may relent: "Just so our inventory doesn't increase as fast as our sales." The controller wants sales to increase faster than inventory. He wants higher turnover because higher turnover means higher return on inventory investment. He doesn't know what the inventory level should be, so he plays the devil's advocate. It's a bargaining game. The inventory manager must know what the inventory level should be.

This chapter tells you how to make an inventory plan part of the company's operating budget. In this chapter the inventory manager gets the company's sales budget from the controller. He looks at the sales budget in relation to past years' customer service and turnover levels and then sets customer service and turnover goals for the future. He converts the customer service and turnover goals into dollars of inventory by month for the future. He reviews this preliminary inventory plan with the controller, the sales manager, and the operations manager. He uses the final inventory plan to control the company's total purchases during the month.

In most companies inventory levels and turnover are emotional subjects, and the inventory manager's performance is measured against rules of thumb. The focus forecasting inventory management system eliminates the emotion by making the inventory plan part of the company's operating budget.

The inventory manager is responsible for one of the major assets in the company. In most companies inventory is 33 percent to 50 percent of total investment. The inventory manager's performance should be measured in the same way that the performance of other company managers is measured. Most companies use budgets to measure how well managers control the company's investment in people and equipment. It's time that companies use the operating budget to measure how well the inventory manager controls the company's investment in inventory.

In this chapter you will see how focus forecasting marries the company's operating budget to its item-buying decisions. This is the only inventory management system I know of that *automatically* marries the company's operating budget to detailed buying decisions.

This chapter describes the first three steps in the inventory management system: the controller budgets total sales, the inventory manager plans total inventory levels, and the inventory manager sets total safety stock levels.

THE CONTROLLER BUDGETS TOTAL SALES

The inventory manager could make his own forecasts of total sales, since he has all of the historic records that the controller has and knows the different techniques for forecasting total sales. But in order to enlist the controller's cooperation, it's better if the inventory manager works from the controller's sales budget. Buyers and inventory planners understand and trust computer forecasts based on their experience. Controllers understand and trust inventory plans based on their forecasted sales budgets.

There is a caution here. Accounting sales budgets generally understate expected sales to hold down expense budgets. What the inventory manager wants from the controller is the company's best estimate of the total actual sales expected in coming months. The inventory manager does not need that sales forecast broken down by division, department, or branch because the inventory management system will determine how to distribute the inventory dollars by division, department, and branch. As a matter of fact, the inventory management system will determine how best to distribute those dollars by individual item. Figure 4.1 is an example of total forecasted sales.

In some companies the controller gets the budgeted sales forecast from the marketing department. In others the president determines

Figure 4.1   TOTAL FORECASTED SALES FOR THE SECOND HALF OF THIS YEAR (THOUSANDS OF DOLLARS)

Month	Total sales
July	10,000
August	8,000
September	7,000
October	10,000
November	10,000
December	10,000

the forecasted sales. In still others the controller forecasts the total sales. Who provides the sales forecast doesn't really matter, as long as everyone is using the same sales budget to plan expenses and investments.

The inventory manager must convert the total sales forecast into sales dollars at cost since all the inventory levels on the financial statement are expressed at cost. To calculate turnover the inventory manager needs both the inventory level and sales at cost. The inventory manager also uses sales at cost to determine how many months on hand are in inventory at the end of a particular month.

If the company's gross profit percentage varies widely from year to year, the inventory manager should get the sales at cost from the controller. If it is fairly consistent year after year, the inventory manager can convert the total sales to sales at cost by subtracting the gross profit percentage from 100 percent. For example, if the gross profit percentage is 10 percent, the cost of goods sold is 90 percent.

The inventory manager can look at the previous year's history to select the gross profit percentage for each month in the coming months and then calculate the cost of goods sold percentage. The cost of goods sold percentage times the controller's budgeted sales forecast gives sales at cost for each month. Figure 4.2 gives an example showing total sales and sales at cost.

If the inventory manager calculates the sales at cost he should review the resulting sales at cost forecast with the controller. The controller may know something that will change the historic gross

Figure 4.2 TOTAL FORECASTED SALES AT COST FOR THE SECOND HALF OF THIS YEAR (THOUSANDS OF DOLLARS)

Month	Total sales ($)	Cost of goods sold (%)	Sales at cost ($)
July	10,000	83	8300
August	8,000	90	7200
September	7,000	90	6300
October	10,000	76	7600
November	10,000	80	8000
December	10,000	91	9100

profit percentage relationship for a given month. More importantly, reviewing the sales-at-cost schedule with the controller shows respect for the controller's judgment. The more the inventory manager uses the controller's judgment in preparing a sales forecast, the more the controller will respect the inventory manager's judgment in determining inventory levels.

The inventory manager should get a new forecast of the coming month's sales every time the controller changes the company's operating budget. Sometimes actual sales differ significantly from the budgeted sales, but the inventory manager should continue to use the controller's forecasted sales unless the controller issues a new budget for the entire company. In this way the inventory manager's plans are tied in with the company's operating budget. Once the inventory manager has a schedule of sales at cost he is ready to plan inventory levels based on historic customer service and turnover.

CUSTOMER SERVICE AND TURNOVER

Customer service and turnover are the keys to future inventory plans. Customer service is lower when turnover is higher, and higher when turnover is lower. The greater the inventory relative to sales, the higher the customer service level will be.

A company can measure customer service in two ways. The first way measures the number of lines a customer receives of the total lines ordered. Let's call this measure the line-fill rate. The second way measures the dollars the customer receives of the total dollars ordered. Let's call this measure the dollar-fill rate.

Suppose you ordered ten lines of merchandise, that is, ten different items. Eight lines on your order are filled, two are canceled. You received $330.05 worth of your $468.05 order. Figure 4.3 shows how we would calculate the line-fill rate and the dollar-fill rate. Your line-fill rate is 80 percent, the lines filled (8) divided by the total lines. Your dollar-fill rate is 71 percent, the dollars received ($330.05) divided by the dollars ordered ($468.05).

Line-fill rate is a better measure of customer service, while dollar-fill rate is a better measure of the inventory control system's ability to earn sales dollars and gross profits for the company. The

Figure 4.3   A SHIPPING ORDER

Ship to		Date 8/17 This year
Harry Saul's Hardware 3 Early Delivery Street Small Town, USA		
Quantity	**Description**	**Amount**
3	Stanley hammers	$ 14.95
144	Cadium swivel snaps	3.45
12	5 lb. tins HTH	10.50
1	Lawn mower	96.80
15	Broiler pans	15.40
~~24~~	Screwdrivers	14.00
9	Escutcheon pins	.45
~~8~~	Bench grinders	124.00
6	3-Speed bicycles	180.00
2	Gallons white paint	8.50
	Total	$468.05

X means your order was canceled

inventory planning in this chapter is based on line-fill rate.

Line-fill rate goals vary by company, by industry, by month. They even vary in the way they are calculated. Some companies "backorder"; they keep a record of canceled customer orders and ship the canceled customer orders when goods become available. The line-fill rate goals in this chapter assume a "no backorder" rule, so a customer must order a canceled item over and over until goods become available. A line-fill rate goal of 92 percent on "no backorder" is usually equal to a line-fill rate goal of 96 percent in a company that backorders. Companies that practice backordering avoid customer reorder of canceled items. Since their customers order fewer canceled items, their fill rate is higher automatically than companies who have "no backorder" rules. Line-fill rate, then, is a measure of customer service, and is the first key to future inventory levels.

Turnover, on the other hand, shows how well inventory is moving through the business. The higher the turnover, the lower the line-fill rate. The classic turnover calculation is annual sales at cost divided by average inventory investment at cost. For example, suppose a company had $60 million annual sales at cost, and the average inventory was $10 million. The turnover would be six.

Now to convert the turnover to months on hand, we first calculate the average month's sales: $60 million annual sales at cost divided by twelve months equals $5 million sales at cost per month. We then divide the average inventory balance by the average monthly sales at cost. In this example $10 million divided by $5 million equals two. So a turnover of six equals two months on hand.

Since inventory is planned to support future sales, turnover should be in terms of future sales. Thus turnover calculations in this chapter are in terms of months on hand, and months on hand are in terms of future sales.

When planning the inventory, the inventory manager is interested in the months on hand at the end of each particular month. Figure 4.4 shows how to calculate the months on hand at the end of May in our example. At the end of May the company has $12 million or 2.2 months on hand. In other words, if the company did not replenish inventory the May 31 inventory balance of $12 million should last for 2.2 months. Step 1 in Figure 4.4 shows that the May 31 inventory balance is high enough to cover June sales. Step 2 shows that it is high enough to cover both June and July sales. Step 3 shows it is high enough to cover June and July sales and 20 percent of August sales. Therefore the May 31 inventory balance is 2.2 months on hand. This is how the inventory manager calculates months on hand for each month of the year.

Figure 4.4 CALCULATING MONTHS ON HAND

May 31 inventory $12 million

Month		June	July	Aug
Sales at cost ($ million)		5	6	5

The May 31 inventory equals 2.2 months on hand

Step 1:	May 31 inventory	$12 million	
	Less sales for June	5 million	1 month on hand
Step 2:	June 30 inventory	$ 7 million	
	Less sales for July	6 million	2 months on hand
Step 3:	July 31 inventory	$ 1 million	
	Divided by sales for August	5 million	2.2 months on hand

The turnover goals in this chapter are in terms of months on hand. Months on hand is the second key to future inventory levels. The inventory manager uses the relationship of past years' months on hand to line-fill rate to plan future inventory levels. This relationship is unique for each month of the year. The next section shows how the inventory manager plans future inventory levels for July, August, and September of the coming year.

THE INVENTORY PLAN

The inventory manager should look at company records for many years to determine the relationship between months on hand and line-fill rate. To make this inventory planning example simple, we will just use last year's history to set goals for line-fill rate and months on hand. Figure 4.5 gives the inventory history for this example for last year.

Figure 4.5 LAST YEAR'S INVENTORY HISTORY (THOUSANDS OF DOLLARS)

Month	Inventory	Sales at cost	Months on hand	Line-fill rate (%)
January	9,000	6200	1.63	86
February	9,700	5300	1.64	88
March	10,000	5800	1.54	90
April	10,200	6100	1.36	92
May	10,500	7200	1.29	90
June	9,200	8300	1.22	87
July	9,000	7700	1.34	85
August	9,300	6900	1.45	85
September	9,800	6100	1.41	85
October	10,000	7100	1.51	87
November	10,300	6500	1.60	88
December	10,600	6800	1.50	83

When the inventory manager prepares an inventory plan, the line-fill rate goals that he sets for August, September, and October determine the months on hand in July, August, and September. The first thing the inventory manager must do is set fill-rate goals for

August, September, and October. It doesn't make any sense to set a line-fill rate goal for July at this point. The July line-fill rate is already determined by the June inventory level.

SETTING LINE-FILL RATE GOALS

Here are ten things the inventory manager should consider in setting line-fill rate goals:

1. What is the company policy on line-fill rate? Is it possible to reach? Has the line-fill rate goal ever been reached while meeting the company turnover goal?
2. How are suppliers shipping? Are goods scarce? Are delivery times getting longer or shorter? Are delivery times inconsistent?
3. How is our recent line-fill rate performance? If it's poor now, can we make it better?
4. What is our competition doing? Is their line-fill rate higher than ours? Are we losing business to competition because their line-fill rate is higher than ours?
5. Are we borrowing money to support inventory now? Are we running out of space?
6. Is our current inventory healthy? Do we have many inventory excesses? Are our months on hand exceeding plan?
7. Do we have new systems planned to improve our customer service/turnover relationship? Are our records accurate? Are our forecasts accurate? Do we really know what's on order?
8. How many items do we have in inventory now versus last year? A lot of new items in inventory mean a high degree of risk. New item demand is uncertain. It takes more inventory to support customer service with a high new item inventory mix.
9. Can our operations department handle changes in the inventory level? Will goods pile up on the receiving dock if we quickly increase inventory? Will we lay off people if we quickly decrease inventory?
10. Are we at the end or beginning of a season? If we are at the end, high line-fill rates mean a high risk of leftovers. If we are at the start of a season, high line-fill rates are safe.

These considerations should go through the inventory manager's mind at the time he is setting line-fill rate goals for the coming period.

Let us suppose that the inventory manager has weighed these considerations and has come up with the line-fill rate goals for August, September, and October shown in Figure 4.6. As the figure shows, the inventory manager has chosen the low line-fill rate goal of 88 percent for August because August is the end of the summer season. He has set higher goals for September and October because they are the beginning of the fall and winter season.

Figure 4.6 FILL RATE (PERCENT)

	Last year's actual	This year's goal
August	85	88
September	85	89
October	87	90

SETTING MONTHS ON HAND GOALS

Once the inventory manager has established line-fill rate goals, he must plan the months on hand at the end of each month to support those line-fill rate goals. A very small increase in months on hand moves the line-fill rate from 85 percent to 88 percent. It takes a greater increase to move from 85 percent to 89 percent. Once the line-fill rate approaches 90 percent, it takes large increases in months on hand to move it up another point.

Inventory planning is scientific gambling; it is risk taking that must follow the laws of probability. Appendix 9 discusses the probability distribution of line-fill rate versus months on hand. For now, though, just note that it takes a large increase in months on hand to move the line-fill rate from 87 percent to 90 percent, and a much smaller increase to move it from 85 percent to 88 percent. Figure 4.7 shows the months on hand versus the line-fill rates for last year.

In the example in figure 4.7 the inventory manager must now plan months on hand for this year to raise August's line-fill rate to 88 percent, September's rate to 89 percent, and October's to 90

Figure 4.7   LAST YEAR'S ACTUAL

	Months on hand		Line-fill rate
July	1.34		
August	1.45	August	85
September	1.41	September	85
October		October	87

percent. The inventory manager compares months on hand versus line-fill rate for a number of years to determine the months on hand in July that will produce an 88 percent line-fill rate in August. The inventory manager goes through the same procedure to set months on hand goals for August and September based on the line-fill rate goals for September and October.

This is not the time to rethink the considerations that went into setting the line-fill rate goals. Now the only concern should be the best estimate of the months on hand that will generate the desired line-fill rates. The inventory manager who keeps the original goals but changes months on hand because of other pressures is planning to be wrong. The only reason to change months on hand is a change in the line-fill rate goal.

Figure 4.8 gives the months on hand plan for July, August, and September of our example. The inventory manager has increased the July months on hand 0.11 months over last year in order to improve the line-fill rate from last year's 85 percent to this year's 88 percent. He has increased the August months on hand 0.20 to get September's line-fill rate of 89 percent and increased the September months on hand 0.40 to get a 90 percent line-fill rate in October.

Figure 4.8   THIS YEAR'S PLAN

	Months on hand		Line-fill rate (%)
July	1.45		
August	1.65	August	88
September	1.81	September	89
October		October	90

This plan shows that the inventory manager knows that a greater increase in months on hand is needed to raise the line-fill rate from 87 percent to 90 percent than to raise it from 85 percent to 88 percent.

The inventory manager plans the months on hand by specific months. The months on hand at the end of July supporting August sales are not the same as the months on hand at the end of August supporting September sales. Each month has its own seasonality and its own sales mix. Each month, year after year, has its own relationship of months on hand to line-fill rate. Now the inventory manager converts the months on hand to inventory dollars.

CONVERTING MONTHS ON HAND TO INVENTORY DOLLARS

Figure 4.9 shows how the inventory manager converts months on hand to dollars. The months on hand must be related to the controller's budgeted sales at cost. In July the planned months on hand was 1.45. The inventory manager converts that planned 1.45

Figure 4.9 CONVERTING MONTHS ON HAND TO INVENTORY DOLLARS
(THOUSANDS OF DOLLARS)

Convert July's 1.45 months on hand to inventory dollars	
1.00 = 100% of August forecasted sales at cost	$ 7,200
+ .45 = 45% of September forecasted sales at cost	2,835
1.45 = July inventory	$10,035
Convert August's 1.65 months on hand to inventory dollars	
1.00 = 100% of September forecasted sales at cost	$ 6,300
⊢ .65 = 65% of October forecasted sales at cost	4,940
1.65 = August inventory	$11,240
Convert September's 1.81 months on hand to inventory dollars	
1.00 = 100% of October forecasted sales at cost	$ 7,600
+ .81 = 81% of November forecasted sales at cost	6,480
1.81 = September inventory	$14,080

months on hand to inventory dollars. He adds 100 percent of August sales to 45 percent of September sales to get the July 31 planned inventory level of $10,035,000. The inventory manager goes through the same calculation for August and September.

THE PRELIMINARY PLAN

The inventory manager gives a copy of the preliminary plan for the next three months' inventory to the controller, the sales manager, and the operations manager. Figure 4.10 shows the preliminary plan for our example.

Figure 4.10 PRELIMINARY INVENTORY PLAN FOR THE THIRD QUARTER OF THIS YEAR (THOUSANDS OF DOLLARS)

This year	Planned inventory[a]	Budgeted sales at cost[b]	Planned months on hand[c]	Line-fill rate
July	10,035	8300	1.45	89[d]
August	11,240	7200	1.65	88[e]
September	14,080	6300	1.81	89[e]
October		7600		90[e]
November		8000		
December		9100		

[a]From planned months on hand and forecasted sales at cost
[b]From the controller
[c]From last year's months on hand versus line-fill rate
[d]Based on June's months on hand
[e]Goals for this year

The controller will work the planned inventory levels into the cash forecast. If the company will have a cash problem in future months, the controller may ask the inventory manager to lower the planned inventory levels. If the controller foresees a cash surplus, he may ask the inventory manager to increase the expected line-fill rate and put more dollars in inventory. This increase will improve the company's sales and gross profits for the period.

The sales manager looks at the planned line-fill rate in relation to what the competition is doing. The sales manager may ask the inventory manager for higher planned line-fill rates.

The operations manager looks at the inventory plan from the standpoint of work. The planned $1,205,000 increase in inventory in August and $2,840,000 increase in September may imply an uneven workload, so the operations manager may ask for a smoother increase in inventory. The operations manager may want the inventory to increase $2 million in August and $2 million in September.

THE FINAL PLAN

The inventory manager must work with the controller, the sales manager, and the operations manager to determine the final inventory plans which will become part of the company's operating budget. Once the inventory plan is part of the operating budget, the inventory manager can be measured in the same way other company managers are measured. Just as the operating budget measures how well other managers control the company's investment in people and equipment, it should also measure how well the inventory manager controls the company's investment in inventory.

The monthly analysis of variances against budget can pinpoint the causes of poor customer service and turnover. With this information, the company can take steps to correct any problems which appear. The inventory plan is thus a necessary part of the operating budget of all companies.

Now that the inventory plan is final, the inventory manager must control how much to buy.

CONTROLLING HOW MUCH TO BUY

Figure 4.11 shows an example of a final inventory plan for the third quarter of this year. As of July 31, there will be $10,035,000 in inventory. The inventory manager plans to have $11,240,000 in

Figure 4.11 FINAL INVENTORY PLAN FOR THE THIRD QUARTER OF THIS YEAR (THOUSANDS OF DOLLARS)

Month	Planned inventory	Budgeted sales at cost	Planned months on hand	Line-fill rate (%)
July	10,035	8300	1.45	89
August	11,240	7200	1.65	88
September	14,080	6300	1.81	89
October		7600		90
November		8000		
December		9100		

inventory at the end of August. To see how much must be added to inventory during August to meet this goal, look at the following calculations.

August 31 inventory on hand	$11,240,000
Add August forecasted sales at cost	7,200,000
Goods available for sale during August	$18,440,000
Subtract July 31 inventory on hand	10,035,000
Required August replenishment	$8,405,000

So the inventory manager must add $8,405,000 of inventory during August to have $11,240,000 of inventory on hand at the end of August. If on the average it takes one month to replenish inventory the inventory manager must purchase $8,405,000 in July.

Now we have the first two steps of this inventory management system. The controller has budgeted the total sales and the inventory manager has planned the total inventory levels. Up to this point this inventory planning approach can be used in any company. The next step is what makes focus forecasting's inventory management system unique. It is different from any other inventory management system I know of.

At the beginning of July the focus forecasting inventory management system prints an inventory position report. This report tells the inventory manager how much will be purchased in July with the existing safety stock settings. It tells the inventory manager how to change the safety stock settings to increase or decrease July purchases. The inventory manager sets total safety stock levels for the coming month. His safety stock level decision controls the total purchases for July. This inventory management system gives an inventory manager the ability to marry financial planning to individual item ordering.

SETTING TOTAL SAFETY STOCK LEVELS

Figure 4.12 is an example of an inventory position report for July 1 of this year. This inventory position report says that the buying system will purchase $7,609,000 at the current safety stock setting. The current safety stock setting is 0.4 to 1.4 months. The report says to increase purchases to $8,219,000; the inventory manager must change the safety stock setting to 0.8 to 1.8 months. The report says to decrease purchases to $6,839,000; the inventory manager must change the safety stock setting to 0.0 to 1.0 months. This safety stock range is the inventory manager's faucet. The inventory manager turns this faucet to control the flow of purchases into the inventory.

Figure 4.12 INVENTORY POSITION REPORT (THOUSANDS OF DOLLARS)

Total Company	July 1
1 Fill-rate	89%
2 Months on hand	1.53
3 Months on order	1.15
4 On-hand dollars	$10,312
5 On-order dollars	$7,905
6 June demand at cost	$6,789
Forecasted demand at cost	
7 July	$6,918
8 August	$6,453
9 September	$7,107
10 October	$7,886
Planned purchases with	
11 0.4 to 1.4 months of safety stock	$7,609
12 0.8 to 1.8 months of safety stock	$8,219
13 0.0 to 1.0 months of safety stock	$6,839
14 Excess inventory dollars	$1,347
15 Average vendor delivery lead time	1.0 months
16 Number of active stockkeeping units	53,843

Remember, the buyers change only 8 percent of the system's buying decisions. The net result of these changes is only a 5 percent difference between suggested and actual purchases. By setting the total safety stocks the inventory manager can control the total purchases for any month within narrow limits. In chapter 5 we will discuss how the safety stock divisions are set from 0.4 to 1.4 in the first place. For now let's look at how the inventory manager determines the exact total safety stock setting for July.

First the inventory manager must determine how much a 0.1 month increase in safety stock will increase the purchases for the month of July. A safety stock setting of 0.4 to 1.4 months will generate purchases of $7,609,000. A safety stock setting of 0.8 to 1.8 months will generate purchases of $8,219,000. By subtracting one from the other you can see that a 0.4 month increase in safety stock will increase purchases $610,000. So to determine how a 0.1 month increase in safety stock will increase purchases, we divide $610,000 by 4 and get $152,500. Thus a 0.1 month increase in safety stock will increase purchases $152,500.

Safety stock of 0.8 to 1.8 months will generate purchases of $8,219,000. The inventory manager has already calculated that he wants purchases of $8,405,000 to meet his inventory plan. The $8,405,000 is an increase in purchases of $186,000 over the 0.8 to 1.8 month safety stock setting. This $186,000 is more than the $152,500 that a 0.1 increase in the safety stock would generate, so he needs an increase of at least 0.1 months. An increase in safety stock of 0.2 months would generate an additional $305,000 worth of purchases. That would be more than the inventory manager wanted. Therefore, the safety stock range should be 0.9 to 1.9 months. This safety stock setting will generate purchases of $8,371,500. This is as close as he can come to generating the planned purchases with a 0.1 month gradation in the safety stock setting.

The inventory manager gives the computer department a total safety stock setting of 0.9 to 1.9 months; it will be used in the calculation of every suggested buy throughout the month of July. The net result of the individual suggested buys for all the items in the company's line will amount to $8,371,500. This is how the inventory manager sets total safety stock levels to generate the purchases needed to meet the inventory plan. Setting the total safety stock levels is the third step in the inventory management system.

The controller has budgeted the total sales for the company; the inventory manager has planned the total inventory levels; and, now, the inventory manager has set total safety stock levels. These levels will produce the total purchases needed to generate the desired inventory levels.

SUMMARY

Inventory goals are a dilemma. Higher customer service means lower turnover. Higher turnover means lower customer service. Without an inventory plan the inventory manager can never win.

The inventory manager uses the controller's sales budget to gain the controller's respect for the inventory plan. We've seen the considerations the inventory manager goes through to set line-fill rate goals. We've seen how he uses past years' monthly turnover versus customer service to set the months on hand needed to meet line-fill rate goals. We've seen how he uses the controller's sales budget to convert these months–on–hand goals to inventory dollars. With the forecasted sales and the planned inventory dollars for a month, the inventory manager can decide the coming months' total purchases.

The focus forecasting inventory management system gives the inventory manager control of the total purchases for the month. The inventory position report lets the inventory manager set total safety stock levels to generate the needed purchases. It's the only system I know of that automatically combines financial planning and item buying. Of course, it works only because the buyers trust the buying system. Remember, the actual purchases that the buyers make vary from the suggested buys by only about 5 percent.

The inventory manager consults fellow managers before the inventory plan becomes part of the operating budget. With the inventory plan as part of the operating budget, the company can steadily improve customer service and turnover. Making inventory management part of the company's management takes the emotion out of inventory management. It's time the inventory plan became a part of all companies' operating budgets!

In the next chapter we will see how to develop logical ordering strategies. We will see how the inventory manager's total safety stock level affects the purchase of every item in the company's line.

Chapter 5

How to Develop Logical Ordering Strategies

So far we have discussed the first three steps in our inventory management system. The controller budgeted the total sales. The inventory manager planned total inventory levels. The inventory manager calculated the planned purchases for the coming months.

On the first of the month focus forecasting projected demand for every item in the company line and calculated planned purchases using its logical ordering strategies. Focus forecasting told the inventory manager what the planned purchases would be at different total safety stock settings, and the inventory manager set total safety stock levels for the coming month. The ordering strategies use these total safety stock settings to calculate suggested buys for every item. At the end of the month the sum of these suggested buys equals the inventory manager's planned purchases.

In this chapter we will see how to calculate a suggested buy. We'll look at reorder point quantity and reorder point time. We'll look at the three parts of reorder point time: review time, lead time, and safety stock time. We'll calculate a suggested buy for a broiler pan. We'll see end-of-season logical ordering strategies.

Some of these terms may sound familiar to you; they are used in many inventory management systems. But don't skip over this chapter. The focus forecasting inventory management system is different. It allows the inventory manager to control *every* part of reorder point time, not just safety stock time.

SUGGESTED BUY

Focus forecasting calculates a suggested buy by subtracting the available quantity from the reorder point quantity. *The available quantity is the quantity on hand plus the quantity on order.* For example, if we have 50 broiler pans on hand, and another 50 on order, we have 100 available. Suppose the reorder point quantity is 300. To determine the suggested buy, we subtract the available quantity (100) from the reorder point quantity (300). Thus the suggested buy is 200 broiler pans.

We review broiler pans once every month. Every time the available quantity is less than the reorder point quantity we suggest a buy for broiler pans. *The suggested buy is the reorder point quantity less the available quantity.* This is different from most other inventory management systems. Other inventory management systems suggest a buy when the available quantity is less than the reorder point quantity. But these other inventory management systems use some economic order quantity as the suggested buy quantity. In focus forecasting the reorder point quantity less the available quantity is already an economic order quantity. In chapter 6 we will see the details of this economic order quantity strategy. For now we just need to know that focus forecasting calculates the suggested buy by subtracting the available quantity from the reorder point quantity.

REORDER POINT

The reorder point in this finished goods inventory management system is not the classic reorder point that randomly launches an order any time during the month. In focus forecasting, *the reorder point quantity is the sum of projected demand over the reorder point time.* Suppose the reorder point time is three months, and the projected demand is 100 units per month. Then the reorder point quantity is 300 units.

The reorder point time is the sum of three lengths of time: review time, lead time, and safety stock time. The review time is the time between each buying decision. If we look at an item once a month for a buying decision, the review time is one month. If we look at an item twice a month for a buying decision, the review time is half a month. If we look at an item every two months for a buying decision, the review time is two months. The lead time is the time it takes goods to get on the shelf after we make a buying decision. The safety stock time is the time that we can wait for a late vendor's order without running out of stock. This same safety stock time covers low forecasting errors. Let's look in more detail at each element of time that makes up the reorder point time.

REVIEW TIME

The review time is the time that passes between buying decisions. Shorter review time means smaller order quantities, so the more often we order, the less we order at one time. Smaller order quantities mean lower inventory investment and so, lower inventory carrying costs.

Longer review time means larger order quantities. Larger order quantities mean the company can more often meet volume and freight discount requirements. Also, larger order quantities mean fewer orders to handle in the course of a year and lower inventory order costs.

The company's total inventory cost is the sum of its inventory carrying costs and its inventory ordering cost. The best review time is the one that results in the least total inventory cost.

In the next chapter we will see a formula that always reduces inventory investment and work. For now let's just look at the review time impact on inventory investment and work.

Look at figure 5.1. For this example we will say that annual sales of this item is $12,000 and that monthly sales are $1000. This way we won't need any safety stock to cover low sales forecasts because we know what the sales will be. In this example the only thing that determines the size of the inventory investment is the review time.

The review time is 1.0 months. That means that we look at this item for a buying decision once a month. The lead time is 1.0 months. That means that it takes goods one month to get on our shelf after we make a buying decision. The beginning on hand on this item is $1000.

Figure 5.1   INVENTORY INVESTMENT VERSUS WORK

Annual Sales: $12,000 ($1,000 every month)
Review time: 1.0 months
Lead time: 1.0 months
Beginning on hand: $1,000
Inventory
Investment ($)

Inventory Investment — $500
Work — 12 units

So now we are at the beginning of month one. The sales are $1000 every month, we have enough inventory on hand to last one month, and it takes one month to get more inventory. So we'd better buy whatever inventory we need for the following month right now. Since the following month's sales will also be $1000, we will buy $1000 worth of inventory right now. During month one our $1000 on hand runs down to zero on hand as we actually sell the $1000 worth of goods during the month. Just at the moment we run out of stock the $1000 that we ordered at the beginning of month one arrives.

At the beginning of month two we go through the same process all over again. Every month we order $1000 worth of goods. At the beginning of every month we start off with $1000 on hand. At the end of the month we wind up with zero dollars on hand. The average inventory investment is $500. The average inventory investment is one-half of one month's supply, half the review time.

Look again at figure 5.1. The units of work are the number of periods in a year divided by the review time period. Since the review time is 1.0 months, there were twelve units of work during the year. That means someone had to make a buying decision twelve times a year. Someone had to receive the order twelve times a year. Someone had to put the order away twelve times a year. Someone had to pay an invoice for the order twelve times a year.

Let's look at the impact of changing review time on inventory and work. On an item with annual sales of $12,000 with a review time of 1.0 months we have an average inventory investment of $500 and work of twelve units. If we change the review time to 0.5 months we cut the average inventory investment in half and double the units of work. On an item with annual sales of $12,000 and a review time of 0.5 months, we have an average inventory investment of $250 and twenty-four units of work. If we change the review time to 2.0 months we double the average inventory investment and halve the units of work. An item with annual sales of $12,000 and a review time of 2.0 months gives an average inventory investment of $1000 and six units of work.

You can see how review time affects inventory size and work. In chapter 6 we will see how the inventory manager can choose review times that reduce both inventory size and work. For now we should just know that focus forecasting lets the inventory manager control every part of reorder point time including review time. Now let's look at the second part of reorder time, lead time.

LEAD TIME

The lead time is the time it takes goods to get on the shelf after we make a buying decision. Controlling lead time is just as important as forecasting sales. Some people think all inventory excesses and shortages are the result of forecasting and ordering errors. They spend hundreds of thousands of dollars researching ways to improve forecast accuracy but ignore the control of lead time. Lead time inconsistencies cause as many inventory excesses and shortages as sales forecasting errors. Let's look again at the inventory investment in our example, in which we know exactly what the future sales will be.

Look at figure 5.2. The actual lead time is the expected lead time. In this case the average inventory is one-half the review time and so the average inventory is $500. There are no inventory excesses and no shortages. Just at the moment the on-hand runs out, our new order of $1000 arrives.

Now look at figure 5.3. The actual lead time is only 0.5 months, but the expected lead time is 1.0 months. Now the average inventory

Figure 5.2 INVENTORY INVESTMENT WITH ACCURATE FORECASTS OF LEAD TIME

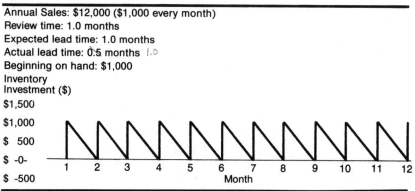

Annual Sales: $12,000 ($1,000 every month)
Review time: 1.0 months
Expected lead time: 1.0 months
Actual lead time: 0.5 months 1.0
Beginning on hand: $1,000
Inventory
Investment ($)

Figure 5.3 INVENTORY INVESTMENT WITH HIGH LEAD-TIME FORECAST ERRORS

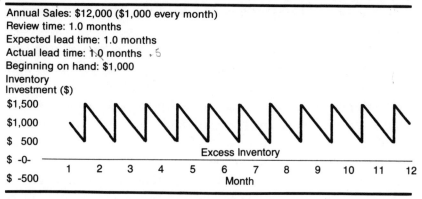

Annual Sales: $12,000 ($1,000 every month)
Review time: 1.0 months
Expected lead time: 1.0 months
Actual lead time: 1.0 months .5
Beginning on hand: $1,000
Inventory
Investment ($)

size has doubled. The average inventory is one-half the review time plus the 0.5 month early delivery, so the average inventory is $1000 and the inventory excess is $500.

In figure 5.4, the actual lead time is 1.5 months but the expected lead time is 1.0 months. The average inventory size remains one-half the review time. The average inventory is now $500. There are shortages of $2000 because the goods arrived on our shelf 0.5 months late. There are excesses because we needed the full $500 average inventory investment to support the $10,000 of annual sales we were able to ship. This same average inventory investment could

Figure 5.4 INVENTORY INVESTMENT LOW LEAD TIME FORECAST ERRORS

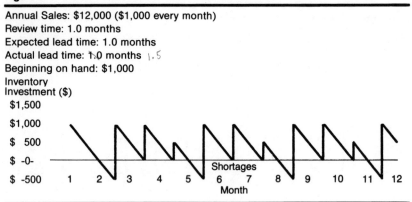

Annual Sales: $12,000 ($1,000 every month)
Review time: 1.0 months
Expected lead time: 1.0 months
Actual lead time: 1.0 months |.5
Beginning on hand: $1,000

have supported $12,000 of annual sales if the actual lead time was consistent with what we expected.

So you can see that controlling lead time is as important as forecasting sales. Now here's the rub. It's impossible to forecast lead time. Lead time on an item is totally dependent on a supplier's performance. Sure, an average of a supplier's lead time performance on all items gives a reasonably accurate forecast of a supplier's future performance on all items. But it is impossible to forecast a supplier's lead time performance on an individual item. Still, lead time is not a random happening; it *is* controllable. A supplier's average lead time can be one month, but our expediters can get almost any item in a week.

The answer is not to forecast lead time. The answer is to *control* lead time. To do so, the inventory manager sets up a distribution requirements plan with major suppliers. He promises the vendor firm purchase order commitments and a plan of future requirements in return for guaranteed delivery and consistent lead time. In chapter 7 you will see the mechanics of this distribution requirements planning system. For now just note the impact of lead time on inventory excesses and shortages.

Now we have seen how review time and lead time affect inventory investment and we can turn to the third part of reorder time: safety stock time.

SAFETY STOCK TIME

The safety stock time is the time we can wait for a late purchase without running out of stock. This same safety stock time will cover low forecasting errors.

We've already seen how the inventory manager sets total safety stock levels to produce the correct total purchases for the month, but how does focus forecasting select a safety stock time for an individual item? Most inventory management systems base safety stock times on forecast error. The more difficult an item is to forecast, the more safety stock time these systems use for the item. Focus forecasting does not base safety stock times on forecast error. Since most forecast errors occur in slow-moving items, strategies that base safety stock on forecast error put most of the safety stock in slow-moving items. This practice is wrong.

Focus forecasting sets safety stock times based on forecasted line demand. Line demand is the number of orders for an item over a period of time. The higher the line demand, the more safety stock time focus forecasting uses for the item. To meet the customer service goal of improving line-fill rate, the inventory manager puts most of the safety stock in the items with the highest line demand. We've seen how the inventory manager sets the total safety stock level. Let's now look at how the inventory manager determines the safety stock range.

First the inventory manager lists all the items in order of descending annual line demand. The item with the highest line demand for the year appears at the top of the report, the item with the lowest line demand appears at the bottom. Look at figure 5.5.

Figure 5.5   ANNUAL ITEM DISTRIBUTION

		June 30 This Year			
Item number	Number of items	Annual line demand	Running total (%)	Annual dollar demand	Running total (%)
08899	1	120,000	1.2	113,026	.1
38421	2	44,826	1.7	34,870	.1
38410	3	34,134	2.0	56,128	.2
17349	4	32,929	2.2	48,743	.3

Item number 08899 is the best-selling item in lines in the company. This item accounts for 1.2 percent of all the lines ordered from the company during the year. This same item is only 0.1 percent of the demand dollars for the year. Items 08899 and 38421 together account for 1.7 percent of all the lines ordered from the company during the year. Together they still account for only 0.1 percent of the demand dollars.

Now the inventory manager divides all the items into six equal parts of annual demand dollars, as figure 5.6 shows. The inventory manager assigns safety stock times to each of the six divisions. Because each division accounts for the same annual demand in dollars, each 0.1 month of safety stock time has the same dollar inventory cost in each division. These divisions allow the inventory manager to put most of the safety stock in the items with the highest line demand, and thus improve line-fill rate without adding to inventory investment or losing dollar-fill rate.

Figure 5.6   COMPANY ANNUAL DEMAND SAFETY STOCK TIME DISTRIBUTION

Annual demand division	Number of items	Minimum demand in lines	Annual demand in 000's of		Safety stock in months
			Lines	Dollars	
I	200	1780	6666	$16,666	1.4
II	800	360	1333	16,666	1.2
III	1,200	120	1000	16,666	1.0
IV	2,800	80	333	16,666	0.8
V	5,000	40	500	16,666	0.6
VI	15,000	0	166	16,666	0.4
Total	25,000		10000	$100,000	

Focus forecasting applies these divisions to every item that it forecasts and buys. An item whose forecasted yearly demand was greater than 360 lines but less than 1780 lines would use the safety stock time for division II. In figure 5.6 division II has a safety stock time of 1.2 months. The inventory manager can try different safety stock times for each division to get the best overall line-fill rate with the same total safety stock time.

Safety stock time is an important decision, because in large measure, safety stock time decides line-fill rate and total inventory investment. All safety stock time adds to inventory investment. Let's look again at the inventory investment in the example with known future sales. Let's see what happens to safety stock when the actual lead time is the expected lead time. We'll use a safety stock time of 1.0 months.

In figure 5.7 the actual lead time is the expected lead time and the actual sales are the forecasted sales. The average inventory investment is one-half the review time plus all the safety stock time. The average inventory is $1500, $500 to cover review time plus $1000 to cover safety stock time.

Figure 5.7 IMPACT OF SAFETY STOCK ON INVENTORY INVESTMENT

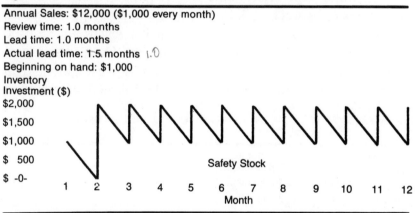

Annual Sales: $12,000 ($1,000 every month)
Review time: 1.0 months
Lead time: 1.0 months
Actual lead time: 1.5 months 1.0
Beginning on hand: $1,000
Inventory
Investment ($)

From this example we can generalize. If the actual lead time equals the expected lead time and the actual sales equal the forecasted sales, the inventory will be one-half the review time plus all the safety stock time. In fact, if the *average* expected lead time and the *average* sales forecast for all items are accurate, the total inventory will be one-half the review time plus all the safety stock time. The length of lead time doesn't matter, as long as it is what we expect. The mix of high and low forecast errors doesn't matter, as long as they are balanced. As a general rule the total months on hand is one-half the review time plus all the safety stock time. Any other inventory investment is simply excess inventory. When the inventory manager sets total safety stock levels, he controls the single biggest part of inventory investment.

Now that we've seen how to determine the reorder time, let's see how focus forecasting calculates a suggested buy for a broiler pan.

EXAMPLE: A BROILER PAN

In chapter 2 we saw how focus forecasting projected demand of 121 broiler pans per month at the beginning of July. Let's look at how focus forecasting would create a suggested buy for that broiler pan and the impact of that suggested buy on months on hand and line-fill rate, month by month.

In figure 5.8, the review time is 1.0 months. That means one month passes between each buying decision. In chapter 6 we will see why the inventory manager selected a review time of 1.0 months, but for now we'll just accept that figure. The lead time is 1.5 months. That means it takes goods 1.5 months to get a broiler pan on the shelf after a buying decision is made. The lead time could be

Figure 5.8 HOW TO BUY A BROILER PAN/JULY 1 THROUGH DECEMBER 1 THIS YEAR

Review time	1.0 months					
Lead time	1.5 months					
Safety stock time	1.2 months					
Reorder point time	3.7 months					
	July	August	September	October	November	December
Current on hand	200	179	206	206	206	206
Current on order	100	148	121	121	121	121
Forecasted demand	121	121	121	121	121	121
Reorder point	448	448	448	448	448	448
Current available	300	327	327	327	327	327
Purchase order	148	121	121	121	121	121

1.7 months, 1.3 months, or whatever; the focus forecasting inventory management system uses the average lead time of all the items that we buy from the broiler pan vendor. The safety stock time is 1.2 months. That means we can wait 1.2 months for a late vendor order without running out of stock. If the vendor's order is on time, we can cover a low forecast error of 145 broiler pans.

The forecast on this broiler pan is 121 units a month. On the average our customers order 3 broiler pans at a time, so the forecasted line demand on this broiler pan is 40 lines a month, or 480 lines a year. The focus forecasting inventory management system selects the division II safety stock setting of 1.2 months for this broiler pan because 480 demand lines are more than 360 lines but less than 1780 lines per year. (See figure 5.6.) The safety stock time is then 1.2 months. If the inventory manager changes the total safety stock level to 0.9 to 1.9 months, the safety stock time for this broiler pan will be 1.7 months.

The reorder point time for this broiler pan is 3.7 months:

Review time	1.0 months
Lead time	1.5 months
Safety stock time	<u>1.2 months</u>
Reorder point time	3.7 months

So the reorder point quantity is 448 broiler pans: 3.7 months times 121 broiler pans per month.

On July 1 there are 200 broiler pans on hand, 100 broiler pans on order, the available quantity is 300 broiler pans. The reorder point quantity is 448 broiler pans. So the broiler pan suggested buy is 148 broiler pans on July 1.

On August 1 there are 179 broiler pans on hand, 148 broiler pans on order, and 327 available. The reorder point quantity is still 448 broiler pans, so the broiler pan suggested buy is 121 broiler pans on August 1.

The suggested purchase will be 121 broiler pans every month from then on. We can see that starting with September 1 on-hand and on-order quantities remain the same month after month. Let's look at the impact on months on hand.

MONTHS ON HAND

First we must determine the high and low points of broiler pan inventory during the month. On September 1 we have 206 pans. Between September 1 and September 15 we sell 61 pans and no pans arrive. Thus the low point is 145 pans on September 15. On September 16 the 121 pans we ordered on August 1 arrive. Thus the high point is 266 pans on September 16.

During September the inventory ranges from a low of 145 pans to a high of 266 pans, so the average inventory investment is 206, an average inventory investment of 1.7 months on hand. We've already seen that one part of inventory investment is one-half the review time, in this case one month. So 0.5 months of the 1.7 months on hand is review time. The other 1.2 months is safety stock time. Remember that the months on hand averages one-half the review time plus all the safety stock time. Any other inventory investment is simply excess inventory.

LINE-FILL RATE

The numbers in this example are static to help you understand how focus forecasting calculates a suggested buy. In the real world in any month many events could change the numbers we are using:

1. The inventory manager may change safety stock time.
2. The inventory manager may change review time.
3. The inventory manager may put the broiler pan vendor on a distribution requirements planning system.
4. The expediters may speed up delivery of broiler pans.
5. The forecasted demand may change.
6. The forecasted demand may be different for each month of the future on seasonal items.
7. The actual demand may be different from the forecasted demand.
8. The actual lead time may be different from the expected lead time.

Inventory management is risk taking, it's scientific gambling on an uncertain future. These logical ordering strategies bet the big safety stock dollars on the company's best-selling items, in lines. In this broiler pan example, the 1.2 months of safety stock protects the line-fill rate. The broiler pan vendor's order can be 1.2 months late or the actual broiler pan demand can be 145 units higher than our forecast before we run out of stock. This 1.2 months' safety stock investment is the price we pay to protect the line-fill rate for broiler pans. In this example this price buys us 100 percent line-fill rate.

In this example the projected demand used to calculate the reorder point quantity started on the date of the buying decision. Now let's see what happens on an end-of-season buying decision by looking at an end-of-season reorder point.

END-OF-SEASON REORDER POINT

In the previous section we went through an example of buying a broiler pan. In figure 5.8 there were 300 broiler pans available at the beginning of the buying period. This section shows the special reorder point consideration for items whose available quantity is less than the demand projected over the lead time. To look at this special consideration let's assume that we had only 50 broiler pans on hand and zero on order at the beginning of July.

From figure 5.9 we can see that only 50 broiler pans are available at the beginning of July. The projected demand over 1.5 months lead time is 121 units a month, or a total of 182 units, so we have fewer broiler pans available than the projected demand over the lead time period. The focus forecasting logical ordering strategies then uses an end-of-season reorder point. To do that the system slides forward in time the length of the lead time. Thus instead of projecting demand over the next 3.7 months from July 1, focus forecasting will project demand over the next 3.7 months from August 15, that is, 1.5 months (the lead time) later. Forecasted demand is 121 broiler pans a month regardless of the month, so the reorder point quantity remains 448 units. Since focus forecasting is using an end-of-season reorder point, it ignores the available in calculating the suggested buys, and calls for the purchase of the reorder point, 448 broiler pans.

When these 448 broiler pans arrive in the middle of August there will be 3.7 months on hand. We've seen that over the long run the

Figure 5.9 HOW TO BUY A BROILER PAN/JULY 1 THROUGH DECEMBER 1
THIS YEAR (END-OF-SEASON REORDER POINT)

Review time	1.0 months
Lead time	1.5 months
Safety stock time	1.2 months
Reorder point time	3.7 months

	July	August	September	October	November	December
Current on hand	50	0	387	266	206	206
Current on order	0	448	0	61	121	121
Forecasted demand	121	121	121	121	121	121
Reorder point	448	448	448	448	448	448
Current available	50	448	387	327	327	327
Purchase order	448	0	61	121	121	121

months on hand is one-half the review time plus the safety stock
time. In the normal reorder point logic the maximum on hand at
any one time is the review time plus the safety stock time, or 2.2
months. With the special end-of-season reorder point the maximum on hand will be the entire reorder point time, 3.7 months.
This extra 1.5 months on hand will cover any extraordinary demand
that occurs because we have run out of stock.

Here is why we expect extraordinary demand. On July 1 we had
only 50 broiler pans on hand. By the second week in July we will
have run out of broiler pans. It takes 1.5 months to get more broiler
pans from our vendor. When our broiler pans run out of stock our
customers will eat into their inventories. By the time more broiler
pans arrive from our vendor in August our customers will have
reduced their inventories significantly and will thus have to replenish their inventory reduction. This is why we expect extraordinary demand when the broiler pan finally arrives from our vendor.

At the beginning of August 448 units are available, more than the
demand forecasted for the lead time, so now focus forecasting
reverts to its normal reorder point logic. Since the reorder point is
448 units and the current available is 448 units, focus forecasting's

suggested buy is zero. On August 15 the 448 units arrive. By September 1, 61 of these units have been sold, leaving 387 units available, so focus forecasting buys another 61 units.

By the beginning of October there are 266 units on hand and 61 on order. Now focus forecasting is back into the normal ordering cycle. The difference between the reorder point of 448 broiler pans and the current available of 327 broiler pans is 121 broiler pans. At the beginning of October focus forecasting buys 121 broiler pans. From that point on it will buy 121 broiler pans each month.

We call this special consideration the end-of-season reorder point. Let's look at how this logic works on a seasonal item, a lawn mower. Look at figure 5.10. The review time, the lead time, the safety stock time, and the reorder point time are the same as they were for the example on the broiler pan. Only the forecasted demand is different. The forecasted demand for the lawn mower is 1200 units in July and 600 units in August. If it weren't for the special end-of-season reorder point logic, focus forecasting would buy 1800 lawn mowers at the beginning of July. By the middle of August all the demand for lawn mowers except for 300 units would have already been past. There would be an excess inventory of 1500 lawn mowers.

The forecasted demand over the lead time period is more than the available on hand; therefore, focus forecasting uses an end-of-season reorder point. Focus forecasting shifts forward the lead time period of 1.5 months. In other words, focus forecasting ignores the demand for July and for one-half of August because it takes 1.5 months to get more lawn mowers from our vendor. When an item is in season it is virtually impossible to expedite delivery, so it is unrealistic to think that we could get more lawn mowers in less than 1.5 months. So focus forecasting ignores the demand during this period. Instead it calculates the forecasted demand for the next 3.7 months from August 15 forward. The forecasted demand over that period is only 300 units, so the reorder point quantity for the lawn mowers is 300.

The current number of lawn mowers on hand is 50. Focus forecasting ignores the current available and buys the entire reorder point quantity. The focus forecasting logical ordering strategies buy

Figure 5.10 HOW TO BUY A LAWN MOWER/JULY 1 THROUGH DECEMBER 1 THIS YEAR (END-OF-SEASON REORDER POINT)

Review time	1.0 months					
Lead time	1.5 months					
Safety stock time	1.2 months					
Reorder point time	3.7 months					
	July	August	September	October	November	December
Current on hand	50	0	0	0	0	0
Current on order	0	300	0	0	0	0
Forecasted demand	1200	600	0	0	0	0
Reorder point	300	300	0	0	0	0
Current available	50	300	0	0	0	0
Purchase order	300	0	0	0	0	0

300 lawn mowers, which arrive August 15. From August 15 through September 1 the lawn mower forecasted sales are 300 units. Theoretically by the end of August there will be zero lawn mowers left. Thus this end-of-season reorder point logic eliminates the excess inventories on seasonal items created by most other inventory management systems.

You should now have a clear understanding of the focus forecasting logical ordering strategies. Just to be sure let's summarize the chapter.

SUMMARY

Focus forecasting calculates a suggested buy for every item in the company's line by subtracting the available quantity from the reorder point quantity. The reorder point quantity is the forecasted demand over the reorder point time. The reorder point time is the review time plus the lead time plus the safety stock time.

The review time is the time between buying decisions. The lead time is the time it takes to get goods on the shelf after a buying decision is made. The safety stock time is the time we can wait for late delivery from a vendor without running out of stock. This same safety stock time will cover low forecast errors if the vendor's goods are delivered on time. Focus forecasting lets the inventory manager control each of these times to meet his inventory objectives.

The inventory manager selects the best review time by using concepts shown in chapter 6. The best review time is the one that reduces the inventory investment and the units of work throughout the company.

We have seen that the inventory manager cannot accurately forecast lead time but can control it by using a distribution requirements planning system. We'll see that system in chapter 7. For now we know that focus forecasting calculates expected lead time for an item by averaging the past lead times for all the items in a vendor's line. Consistent lead times are more important than short lead times.

The inventory manager sets total safety stock levels as part of the inventory planning process. We've seen how he uses the six divisions of safety stock time to concentrate safety stock time in the best-selling items.

The inventory investment is one-half the review time plus all the safety stock time. Any other inventory investment is simply excess inventory. In the example of a broiler pan we have seen how these logical ordering strategies will maintain months on hand equal to one half the review time plus the safety stock time.

If the available quantity for an item is less than the projected demand over the lead time, focus forecasting uses an end-of-season reorder point strategy, which carries extra stock to cover unexpected increases in demand on an out-of-stock item. This same strategy reduces the risk of excess inventories on seasonal items at the end of their season.

We have now seen the first three steps in this inventory management system. The controller has budgeted the total sales. The inventory manager has planned the total inventory levels, based on the controller's total sales budget. The inventory manager has set the total safety stock levels at the beginning of the month. And we have seen the logical ordering strategies that are part of the fourth step in the focus forecasting inventory management system: the buyers buy the individual items in the company's line. The inventory manager controls how much the buyers will purchase by controlling safety stock time, review time, and lead time. In chapter 4 we've learned how he controls safety stock time. In chapter 6 we'll see how he controls review time.

Chapter 6

New Concepts of EOQ

The buyers purchase the individual items in the company's line. Although they are free to change any of the computer's suggested buys, they don't because they trust the system. Because they don't, the inventory manager can control the total inventory size by changing the factors used in the logical ordering strategies to suggest buys for individual items.

The factors that determine the reorder point time in the logical ordering strategies are safety stock time, lead time, and review time. In chapters 4 and 5 we've seen how the inventory manager sets the safety stock time to trade lower turnover for higher customer service. In chapter 7 we will see how the inventory manager trades firm long-term purchase commitments for consistent lead time and guaranteed delivery. In this chapter we will see how the inventory manager sets the review time to trade lower turnover for work reduction throughout the company. Review time is the key to this new concept of economic order quantities. The ideal review time will reduce the total cost of carrying inventory plus the total cost of ordering inventory.

The economic order quantities concept in this chapter is based on finding the ideal review time. What you will see in this chapter is a new concept of economic order quantities, a formula that always reduces inventory investment and the total cost of acquiring inventory. The inventory manager uses this formula in conjunction with a computer simulation process. By changing the parameters in this simulation the inventory manager can achieve one of four different objectives.

1. The inventory manager can reduce both the total cost of carrying inventory and the cost of acquiring inventory.
2. The inventory manager can keep the cost of carrying inventory constant and significantly reduce the cost of acquiring inventory.
3. The inventory manager can keep the cost of acquiring inventory constant and significantly reduce the cost of carrying inventory.
4. The inventory manager can spend all the potential savings for better customer service.

The real advantage of this formula is that it does not require any estimates of the actual costs of carrying or acquiring inventory. Focus forecasting uses this formula to calculate the ideal review time in the inventory management system.

In this chapter you will see the classic economic order quantity formulas and their limitations. You will see how the ideal review time concept of economic order quantities was developed. You will see an example of the ideal review time formula in action on groups of items. You will see how focus forecasting's inventory management system uses the ideal review time to reduce the cost of carrying inventory and the cost of acquiring inventory. My good friend Dr. Joseph Bowman from Carnegie Mellon calls this economic ordering strategy "a mathematical anomaly." You will see that it is a common-sense approach to economic order quantities. It's a simple tool and it works.

First let's look at the classic EOQ formulas and their limitations.

ECONOMIC ORDER QUANTITIES

An order quantity is the number of units a company decides to buy or make when an inventory needs replenishment. In a wholesaling or retailing company the order quantity is the purchase order quantity. In a manufacturing company the order quantity is the production work order quantity. The higher the order quantity, the fewer orders are needed in a year. The lower the order quantity, the more orders are needed in a year. The higher the order quantity, the higher the inventory and the resulting inventory carrying costs. The lower the order quantity, the more orders and the greater the ordering costs. So inventory managers want small order quantities and low inventory carrying costs. Production and warehousing managers want larger order quantities and lower ordering costs.

Economic order quantity formulas have been around so long that they seem to be just academic exercises. You have to stand on a production floor to experience the emotions of economic order quantities. Last year when I visited Black & Decker, I saw a machine that is a miracle of modern-day production. The machine reminded me of the song "Proud Mary."

Proud Mary is a five-hundred-ton stamping machine. Half her body is buried in concrete twenty feet into the ground. A shiny ribbon of rolled steel flows into her grippers. Ka-bum, ka-bum, the stamping arm rises to the ceiling and strikes the steel. Pa-chink, pa-chink, a perfectly formed grass shear fork drops into a basket. When Proud Mary gets indigestion a young man stops the machine, reaches under the lifted stamping arm, and clears the bits of twisted steel. Then Proud Mary starts again with a ponderous rhythm that shakes the whole plant. Ka-bum, pa-chink, ka-bum, pa-chink—4 forks a minute, 240 forks an hour. A million-dollar machine paying its way making a $2 grass shear fork blade twenty-four hours a day, three shifts, six days a week.

When enough forks have been made, Proud Mary makes saw blades. She sits idle, hissing impatience until the old steel is rolled and put away and a different steel is threaded into her grippers. The minutes tick by, the fork dies are unfastened part by part. New dies are assembled. The minutes turn to hours. Each worker hurries. Every hour that Proud Mary sits idle dollars are lost. Proud Mary knows all about economic order quantities. The bigger the order quantities, the more forks or saws Proud Mary can make in a day. The smaller the order quantities, the longer Proud Mary sits around and hisses.

An inventory manager back at headquarters wants Proud Mary to run small orders. The fewer forks committed to inventory at one time and the more frequent the orders, the better inventory turnover will be. The more frequent the orders, the lower the inventory carrying costs. Suppose the inventory manager needed 10,000 loads of forks per year. Let's see what happens in three cases: Proud Mary runs all 10,000 loads of forks in one run, she runs 5000 loads of forks in each run, and she runs 1000 loads of forks in each run. Let's look at the total ordering costs and the total inventory carrying costs for these three cases (fig. 6.1).

Figure 6.1 PROUD MARY ORDERING COST (PRODUCTION NEEDED: 10,000 LOADS OF FORKS)

	Order quantities		
	10,000	5000	1000
Number of orders needed	1	2	10
Paperwork to put orders in production	$1	$2	$10
Cost to pull raw material from stock	8	16	80
Accounting for the movement of order through production	5	10	50
Tearing down and setting up Proud Mary	125	250	1250
Total cost	$139	$278	$1390

Figure 6.1 shows that every time the inventory manager orders a load of forks there is cost involved. There's the cost of the paperwork to put the order into production. There's the cost of pulling the raw material from stock. There's the cost of accounting for the movement of the order through production. There's the big cost of tearing down and setting up Proud Mary. Every order incurs this cost. If the inventory manager places ten orders during the year it theoretically costs ten times as much as ordering one time during the year.

But there are costs for carrying large quantities of forks in inventory, as figure 6.2 shows. If Proud Mary runs 10,000 loads of forks all in one run, the average annual inventory size is 5000 loads of forks. There would be 10,000 loads of forks at the beginning of

Figure 6.2 PROUD MARY INVENTORY CARRYING COST (PRODUCTION NEEDED: 10,000 LOADS OF FORKS)

	Order quantities		
	10,000	5000	1000
Average annual inventory size	5000	2500	500
Interest charges on money tied up in fork inventory	$800	$400	$ 80
State tax on inventory investment	300	150	30
Insurance against fire, theft, flood	100	50	10
Increased risk of obsolescence and deterioration	120	60	12
Other costs—handling, storage, rotation, spoilage, physical counting	100	50	10
Total costs	$1420	$710	$142

the year and zero loads of forks at the end of the year, an average of 5000 loads of forks. If the company is borrowing money to support its inventory, it is paying interest charges on the money tied up in the fork investment. If the company is located in a state that charges tax on inventory investment it pays a tax directly related to the size of the average inventory investment. If the company insures its inventory it pays for insurance against fire, theft, and floods. Grass shears are a relatively new invention and right now they are really popular. The more forks in inventory the greater the risk (and cost) of obsolescence and deterioration.

There are other costs related to the size of the inventory. A warehouse manager probably cannot let 10,000 loads of forks just lay in one spot for the whole year. They have to be moved about to arrange the space, so there's a handling cost. If the warehouse space is rented or if additional warehouse space must be constructed there's a cost for storing the 10,000 loads of forks. If the forks are packaged there's probably a cost for rotating stock to make sure that the oldest boxes are moved first. If there's any moisture in the warehouses the boxes could deteriorate or the forks could even rust. Finally, the more forks in inventory, the more it costs to physically count them for financial purposes.

All these costs are directly related to the average annual inventory size. Theoretically if the order quantity is cut in half the cost of carrying the inventory is cut in half. Ordering loads of forks ten times a year instead of once a year theoretically cuts inventory carrying costs to one-tenth what they would be by ordering all the loads at once. Let's look at the total costs for order quantities of 10,000, 5000, and 1000 loads of forks.

Look at figure 6.3. You can see that an order quantity of 10,000 loads has low ordering costs and high carrying costs. An order quantity of 1000 loads has high ordering costs and low carrying costs. By adding the ordering cost to the carrying cost we can see the total costs of the various order quantity sizes. You can see that somewhere between an order quantity of 10,000 and 1000 is an order size that gives the minimum cost. This quantity size is called the economic order quantity.

Figure 6.3 TOTAL COSTS (ANNUAL PRODUCTION NEEDED: 10,000 LOADS OF FORKS)

	Order quantities		
	10,000	5000	1000
Ordering costs	$139	$278	$1390
Inventory carrying costs	1420	710	142
Total costs	$1559	$988	$1532

For years people have been using the classic economic order quantity formula to determine the order quantity that minimizes the total cost. Appendix 11 shows the derivation of the classic economic order quantity formula. For now let's just use this formula to find the ideal order quantity for running loads of forks on Proud Mary.

Figure 6.4 shows the classic economic order quantity formula in step 1. In step 2 we have substituted the estimated costs for loads of forks into the formula, and in step 3 we have gone through the arithmetic to solve the formula. So theoretically the ideal order size for loads of forks is 3129 loads.

Before we go into the limitations of this classic formula, let's make sure that our arithmetic is correct.

Figure 6.4 CLASSIC ECONOMIC ORDER QUANTITY FORMULA

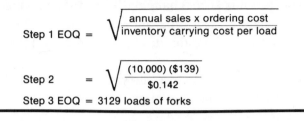

$$\text{Step 1 EOQ} = \sqrt{\frac{\text{annual sales} \times \text{ordering cost}}{\text{inventory carrying cost per load}}}$$

$$\text{Step 2} = \sqrt{\frac{(10,000)\,(\$139)}{\$0.142}}$$

Step 3 EOQ = 3129 loads of forks

Look at figure 6.5. You can see that the total cost for an order quantity of 3000 loads is $889.33. For an order quantity of 3200 the total cost is $888.78. If we move away from the economic order quantity of 3129 loads of forks in either direction the total cost becomes higher. So theoretically 3129 loads of forks is the economic order quantity.

Figure 6.5 CLASSIC EOQ PROOF (PRODUCTION NEEDED: 10,000 LOADS OF FORKS)

	Order quantities		
	3000	3129	3200
Number of orders needed	3.3	3.2	3.1
Average annual inventory size	1500	1565	1600
Ordering cost	$463.33	$444.23	$434.38
Inventory carrying cost	426.00	444.32	454.40
Total cost	$889.33	$888.55	$888.78

LIMITATIONS TO CLASSIC EOQ

There are two major limitations in the classic economic order quantity formula:

1. It depends on estimates of total inventory costs. These estimates are inaccurate.
2. It assumes that total inventory costs are linear. They are not.

Companies estimate the cost of ordering inventory because it is impossible to calculate the cost of ordering every item. Companies have hundreds and thousands of items. The manufacturing process keeps changing. The item mix keeps changing. The production load keeps changing. Calculating exact ordering cost would cost more than it would save. So companies estimate ordering costs, and these estimates are inaccurate.

Companies also estimate the cost of carrying inventory, again because it is impossible to calculate the carrying cost for every item. Some items have more risk of obsolescence and deterioration. Some items are bulky. Some items are hard to count. So again, calculating exact carrying cost would cost more than it would save, and again companies estimate carrying costs and these estimates are inaccurate.

When companies assume that the cost of ordering inventory is linear, they assume that one additional production order costs one additional unit of work. But the cost of ordering is not linear; it depends on the percentage of capacity the company is using. How much does an additional order cost if the computer time is available to print the order? How much does an additional order cost if the raw material people are idle? How much does an additional order cost if the accounting for the movement through production is mechanized? How much does an additional order cost if Proud Mary is sitting idle? So you can see that far from being linear, ordering costs depend on capacity.

Companies assume the cost of carrying inventory is linear, that one additional unit of inventory costs one additional unit of carrying cost. But the cost of carrying inventory, like the cost of ordering inventory, is not linear. It depends on the size of the company's total inventory. Will one additional unit of inventory force us to borrow money? Will one additional unit of inventory force us to get more warehouse space? Will one additional unit of inventory increase our taxes or our insurance premiums? It all depends on the size of our total inventory, and so inventory carrying costs are not linear.

The classic economic order quantity formula looks beautiful but it rarely yields the ideal order quantity. The new concept of EOQs you will see in this chapter does not require estimates of ordering costs or inventory carrying costs. The new concept you will see in this chapter always reduces ordering costs and inventory carrying costs.

We have just seen an example of how a manufacturer uses a classic economic order quantity formula. Let's look for just a minute at the classic economic order quantity formulas in relation to wholesalers and retailers.

WHOLESALE AND RETAIL EOQs

Wholesalers and retailers are using item economic order quantity formulas. The wholesale and retail trade journals are full of such formulas and strategies. But wholesalers and retailers don't write purchase orders for one item. They don't pull raw material for production of an item. They don't account for movement of goods through production for one item, and they certainly don't set up and tear down Proud Marys to produce one item.

There are ordering costs in wholesaling and retailing, but they are related to the total purchase order and the number of items on the purchase order. Wholesaler and retailer ordering costs are related to the number of times an order is placed, not how many times a particular item was ordered. Wholesalers and retailers have the same inventory carrying costs that manufacturers have, but in manufacturing the ordering cost is related to individual items whereas in wholesaling and retailing the ordering cost is related to the number of and the size of the orders from a vendor.

From figure 6.6 you can see that the unit of work in ordering costs is a purchase order, not an item. The computer prints the paperwork for all the items from a vendor. The buyer reviews all the items from a vendor for a buying decision. The trucking company is concerned with the total of the purchase order, not the quantity of any individual item. The volume discounts are based on the total purchase order size, not the quantity of any individual item. The warehouse receiving work is related to the number of shipments received from a vendor and the number of items on the receiving document. Warehouse receiving work is little affected by the quantity ordered on any individual item. The accounts payable clerk is concerned with the number of invoices received from the vendor and the number of items on the invoice. The accounts

Figure 6.6 BLACK & DECKER ORDERING COSTS
(ANNUAL, PURCHASES NEEDED: $20,000)

	Purchase order size		
	$20,000	$10,000	$2000
Number of orders needed	1	2	10
Paperwork and buyer time required to place a purchase order	$50	$100	$500
Full truckload shipping costs vs LTL*	0	200	1000
Lost volume discounts based on purchase order size	0	100	500
Warehouse receiving work to stock and record a purchase order	20	40	200
Matching vendor invoices, shortage claims, paying bills	10	20	100
Total cost	$80	$460	$2300

*LTL = Less than truckload

payable clerk could care less about the quantity of any individual item on the invoice. Thus the cost of ordering in a wholesaling or retailing company is directly related to the number of purchase orders issued to a vendor and the number of items on those purchase orders.

The inventory carrying cost in a wholesaling or retailing company, on the other hand, is exactly the same as the inventory carrying cost in a manufacturing company. Figure 6.7 shows that each element of inventory carrying cost in a wholesaling or retailing company is exactly the same as an element of inventory carrying cost in a manufacturing company.

Now let's look at the total cost in a wholesaling or retailing company. Look at figure 6.8. You can see that the ideal purchase order size is somewhere in between $20,000 and $2000. We could go through the classic economic order quantity formula to determine the ideal purchase order size, but we would be using a formula that needs estimates of average ordering and inventory carrying costs. We know that these estimates are inaccurate, and we know that the ordering and inventory carrying costs are not linearly related to the order size.

Let's look at a new concept of economic order quantity, one that

Figure 6.7   BLACK & DECKER INVENTORY CARRYING COST
(ANNUAL PURCHASES NEEDED: $20,000)

	Purchase order size		
	$20,000	$10,000	$2000
Average annual inventory size	$10,000	$ 5,000	$1000
Interest charges on money tied up in purchase order investment	800	400	80
State tax on inventory investment	300	150	30
Insurance against fire, theft, flood	100	50	10
Increased risk of obsolescence and deterioration	120	60	12
Other costs—handling, storage, rotation, spoilage, physical counting	100	50	10
Total costs	$ 1,420	$ 710	$ 142

Figure 6.8   BLACK & DECKER TOTAL INVENTORY COSTS
(ANNUAL PURCHASES NEEDED: $20,000)

	Purchase order size		
	$20,000	$10,000	$2000
Ordering costs	$ 80	$ 460	$2300
Inventory carrying costs	1420	710	142
Total cost	$1500	$1170	$2442

does not need estimates of ordering and inventory carrying costs. This concept works within the existing company capacity. It never increases work or inventory above the existing level. It does not increase the need for hiring, equipment, space, or borrowing. It does not depend on a linear relationship between order size and inventory cost.

NEW CONCEPT OF EOQ

This new concept of EOQ is based on four observations of all companies' ordering costs and inventory carrying costs:

1. Every company has a review time. A review time is the length of time that passes between buying decisions. The more frequently a company makes a buying decision, the more pur-

chase orders there will be. The more purchase orders there are the higher the ordering costs will be. The less frequently a company makes a buying decision, the more inventory they will have to carry. The higher the inventory the higher the inventory carrying costs will be.

2. The ordering cost in a company is directly related to the lines of work that the company performs in a year. The lines of work that a company performs in a year is the number of times they review an item during the year times the number of items in the company's line.

3. The cost of carrying inventory is directly related to the size of the company's inventory investment. The less frequently a company makes a buying decision the higher their inventory investment and the carrying costs will be.

4. In wholesaling and retailing companies the cost of freight and the cost of lost volume discounts are part of the ordering cost. The less frequently a company makes buying decisions, the more frequently they can meet freight restrictions and get volume discounts.

Let's look at these four observations as they relate to a very small company, a wholesaler who purchases $280,000 a year. The company has twenty-five items in its line. It purchases the twenty-five items from two vendors, vendor A and vendor B. The company reviews its twenty-five items twelve times a year. The vendors are located nearby so there are no minimum freight restrictions on purchase orders. We will look at the company's ordering costs in terms of total lines of work. We will look at the company's inventory carrying costs as some percentage x of the average inventory.

Look at figure 6.9. Let's first calculate the wholesaler's ordering costs in terms of total lines of work. The company orders from vendor A twelve times a year. Vendor A carries one item so the company performs 12 lines of work in a year for vendor A. The company orders from vendor B twelve times a year. Vendor B carries twenty-four items so the company does 288 lines of work for vendor B in a year. The total lines of work for vendors A and B combined are 300 lines of work per year. You can see that this is a

Figure 6.9 A SMALL WHOLESALER

Vendor	Number of items	Review frequency	Annual volume	Minimum freight restrictions
A	1	12	$240,000	none
B	24	12	40,000	none
Total 2	25		$280,000	

Ordering cost	= 300 lines of work
Inventory carrying cost	= x times $11,667 average inventory
Total cost	= 600 lines of work

direct result of the company's decision to review these vendors twelve times a year. If the company reviewed these vendors twenty-four times a year, the lines of work would be doubled. If the company reviewed these vendors six times a year, the lines of work would be halved.

Now look at the company's inventory carrying costs. The company buys $280,000 a year. Each time it reviews a vendor for a buying decision, it buys one-twelfth of this annual volume. Each time the company orders, it orders on the average $23,333 worth of goods. At the beginning of the month it has an inventory of $23,333 worth of goods. At the end of a month theoretically it will have zero goods on hand. So the average inventory investment is $11,667. The company's inventory carrying cost is some percentage x times an average inventory investment of $11,667. If the company orders from these vendors twenty-four times a year it cuts its average inventory carrying cost in half. If the company orders from these vendors six times a year it doubles its average inventory carrying cost.

What happens if the company reviews vendor A twice as frequently and vendor B half as frequently? Figure 6.10 shows that varying the review frequency of the two vendors reduces the lines of work from 300 to 168 and the average inventory from $11,667 to $8333.

Look what happens. When we double the review frequency on vendor A we add only 12 lines of work because vendor A has only one item. When we double the review frequency on vendor A we decrease the average inventory investment $5000. When we de-

Figure 6.10 A SMALL WHOLESALER

Vendor	Number of items	Review frequency	Annual volume	Minimum freight restrictions
A	1	24	$240,000	none
B	24	6	40,000	none

Ordering cost	= 168 lines of work
Inventory carrying cost	= x times $8333 average inventory
Total cost	= 382 lines of work

crease the review frequency on vendor B to six times a year we eliminate 144 lines of work. When we reduce the review frequency of vendor B to six times a year we increase the inventory investment only $1667. The net result is that we reduce the ordering cost 132 lines and the inventory carrying cost x percent of $3334.

This simple exercise makes you wonder whether there is some ideal review frequency mix that reduces the ordering cost in terms of lines of work and reduces the inventory carrying cost by some percentage x times the average inventory investment.

There is. Here is the formula for calculating the ideal review frequency.

$$\text{economic review frequency} = \sqrt{\frac{(0.5)\,(x)\,(\text{annual volume})}{\text{number of items}}}$$

The derivation of this formula is shown in appendix 12. The first step in using it is to calculate the inventory carrying cost as some percentage x of the average inventory investment. That is,

inventory carrying cost$-(x)$(average investment inventory)

The secret to determining x is to look at the current relationship of lines of work to average inventory investment. In our example, the small wholesaler does 300 lines of work and has an average inventory investment of $11,667. That means that it "costs" the wholesaler 300 lines of work to maintain an average inventory of $11,667. So the inventory carrying cost is 300 lines of work. Now what percentage x of this average inventory investment gives this carrying cost? To find x, we solve for it in the following equation

$$x = \frac{\text{inventory carrying cost}}{\text{average inventory investment}}$$

and substitute the values from our example. Thus

$$x = \frac{\dfrac{300 \text{ lines of work}}{\$11,667}}{} = 2.57 \text{ percent}$$

Thus the inventory carrying cost is 2.57 percent of the average inventory investment.

Once we have expressed carrying costs in terms of lines of work, we can express the company's total cost in terms of lines of work. The total cost equals the ordering cost, 300 lines of work, plus the inventory carrying cost 2.57 percent of an average inventory investment of $11,667, or 300 lines of work. Thus the total cost equals 600 lines of work.

We now have a common denominator for expressing ordering costs and inventory carrying costs—lines of work. Let's see what happens when we put these ordering costs and inventory carrying costs into the formula for economic review frequency. We will calculate the economic review frequency for vendors A and B. First we calculate the economic review frequency for vendor A. Look at figure 6.11. In step 1 you see the formula for calculating the

Figure 6.11　A NEW CONCEPT OF EOQ: VENDOR A

Step 1
economic review frequency $= \sqrt{\dfrac{(0.5)\,(X)\,(\text{annual volume})}{\text{number of Items}}}$

Step 2 $= \sqrt{\dfrac{(0.5)(0.0257)(240,000)}{1}}$

Step 3 $=$　55.53 times a year

economic review frequency. In step 2 we put in the values for the percentage x, annual volume, and the number of items for vendor A. In step 3 we carry out the arithmetic, and we see that the economic review frequency for vendor A is 55.53 times a year.

Now let's calculate the economic review frequency for vendor B. Look at figure 6.12. Again step 1 gives the formula for calculating the economic review frequency, step 2 shows the actual values for vendor B, and step 3 shows that the economic review frequency for vendor B is 4.63 times a year.

Figure 6.12   A NEW CONCEPT OF EOQ: VENDOR B

Step 1

$$\text{economic review frequency} = \sqrt{\frac{(0.5)\,(X)\,(\text{annual volume})}{\text{number of items}}}$$

$$\text{Step 2} = \sqrt{\frac{(0.5)(0.0257)(40,000)}{24}}$$

$$\text{Step 3} = 4.63 \text{ times a year}$$

This same formula works for a $200,000,000 company with 2000 suppliers. Let's check the arithmetic for this small wholesaler. Look at figure 6.13. With these economic review frequencies we have reduced the total ordering cost to 167 lines of work per year, a reduction of 44 percent. We have reduced the average inventory investment to $6481 per year, a reduction of 44 percent. Notice that this new concept of EOQ does not require estimates of the actual ordering costs and the actual inventory carrying costs. Notice that the ordering cost and the inventory carrying cost were both reduced 44 percent.

This economic order quantity formula always reduces ordering costs and inventory carrying costs. Any formula that reduces both ordering costs and inventory carrying costs has to be working correctly. The application of this formula automatically reduces work throughout the company and automatically reduces the inventory investment throughout the company. Notice also that the lines of work in relation to average inventory investment has remained the same; 167 lines of work is still 2.57 percent of $6481 average inventory investment.

At this point the inventory manager of the small wholesaling company could use these review frequencies to reduce all the work

Figure 6.13 PROOF OF A NEW CONCEPT OF EOQ

Vendor name	Number of items	Review frequency	Annual volume	Minimum freight restrictions
A	1	55.53	$240,000	None
B	24	4.63	40,000	None

Ordering cost = Number of items × the review frequency

Vendor A	1	×	55.53 = 56
Vendor B	24	×	4.63 = 111
Total lines of work			167

Inventory carrying cost = Annual volume ÷ (2 × review frequency)

Vendor A	$240,000	÷ (2 × 55.53) =	$2,161
Vendor B	40,000	÷ (2 × 4.63) =	4,320

Ordering cost = 167 lines of
work per year

Inventory carrying cost = 2.57 percent
times $6481
average inventory
= 167 lines of
work

Total cost = 334 lines of
work

in the company and to reduce the company's average inventory investment. The real beauty of this economic review frequency formula is that the inventory manager now has four options:

1. He can use the economic review frequencies as they have just been calculated to reduce all the work in the company and the company's inventory investment.
2. He can further reduce the lines of work in the company by reducing the percentage x in the formula.
3. He can further reduce the company's inventory investment by increasing the percentage x in the formula.
4. He can use all the inventory investment savings for extra safety stock investment. In other words, he can trade off the inventory investment savings for improved customer service.

In this example we developed a percentage x for the whole company. A larger company would develop the percentage x for a department or a group of similar items. This insures that when we vary the review frequency we are trading off equal lines of work. It insures that we are trading off equal inventory carrying costs.

This economic review frequency formula has two major advantages over classic EOQ formulas:

1. It does not require estimates of inventory ordering costs and inventory carrying costs. Inventory ordering costs are expressed in terms of lines of work, and inventory carrying costs are expressed as a ratio of lines of work to average inventory investment.
2. It gives the inventory manager tremendous flexibility. If money is short and there is idle capacity the inventory manager can trade off lines of work for inventory investment. If the operations people in the company are working overtime and there is too little inventory he can trade off inventory investment for fewer lines of work. If money is tight and the operations people are working overtime he can still trade off inventory investment for increased safety stock and improved customer service.

This is definitely a new concept of economic order quantities. It gives the inventory manager complete control of one of the key elements in inventory investment. We have seen that the reorder point time is the safety stock time, the review time and the lead time. We have already seen how the inventory manager controls the safety stock time through inventory planning. Now you see how the inventory manager controls review time through this new concept of economic order quantities. You've seen the inventory manager calculate the percentage x, the current ratio of the lines of work to average inventory investment in the company. The inventory manager uses the percentage x in the economic review frequency formula to calculate the ideal review time for each vendor. The ideal review time for each vendor equally decreases the lines of work and the average inventory investment in the company. This formula equally reduces the company's ordering costs and the company's inventory carrying costs, and gives the inventory manager the flexibility to trade off lines of work for inventory investment.

In this section we assumed that the vendors who supply this small wholesaler were located nearby and ignored minimum freight restrictions and volume discounts. Let's see what happens if vendor A has a $10,000 minimum freight restriction.

FREIGHT RESTRICTIONS AND VOLUME DISCOUNTS

Some suppliers are located quite a distance from their customers. The cost of shipping full truckloads differs considerably from that of partial loads. Some suppliers give quantity discounts for larger orders. Earning these discounts can have a great impact on a company's gross profit.

In the last section we saw how the economic review frequency formula came up with an ideal review time that reduces the inventory ordering and carrying costs. This economic review frequency formula is part of a computer calculation that considers volume discounts and freight restrictions.

Let's look at the example of a small wholesaler again. This time let's assume that vendor A is located far from the small wholesaler. Let's assume that vendor A prepays freight if the small wholesaler buys $10,000 each time the small wholesaler orders.

Look at figure 6.14. The 300 lines of work are still 2.57 percent of the $11,667 average inventory investment, x is still 2.57 percent in the formula for calculating economic review frequencies. The only difference is that now there is a $10,000 freight restriction on vendor

Figure 6.14 A SMALL WHOLESALER

Vendor name	Number of items	Review frequency	Annual volume	Minimum freight restrictions
A	1	12	$240,000	$10,000
B	24	12	40,000	none

Ordering cost	= 300 lines of work
Inventory carrying cost	= 2.57 percent times $11,667 average inventory
Total cost	= 600 lines of work

A. At the present time the small wholesaler is having no problem meeting this freight restriction because he buys from vendor A twelve times a year. Each time he buys from vendor A he is buying $20,000, so he is buying more than the freight restriction.

But this freight restriction limits the number of times the small wholesaler can buy from vendor A. To find out how often the small wholesaler can buy from vendor A, divide the freight restriction into the volume. This calculation shows that the greatest number of times the small wholesaler can buy from vendor A and still meet the minimum freight restriction is twenty-four times a year. In the calculation of the economic review frequency, this is the upper limit on how frequently the small wholesaler can buy from vendor A. Let's look at the economic review frequency for the small wholesaler with this freight restriction consideration.

In figure 6.15, you can see that the economic review frequency formula calculated a review frequency of only twenty-four times a year for vendor A. The review frequency was limited to twenty-four times a year because the freight restriction was $10,000. If the small

Figure 6.15   A SMALL WHOLESALER

Vendor name	Number of items	Review frequency	Annual volume	Minimum freight restrictions
A	1	24.00	$240,000	$10,000
B	24	4.63	40,000	none
Ordering cost	= 135 lines of work			
Inventory carrying cost	= 2.57 percent times $9320 average inventory			
Total cost	= 375 lines of work			

wholesaler bought from vendor A any more frequently he would not be able to meet the freight restriction of $10,000. The economic review frequency for vendor B was still 4.63 times a year since there was no freight restriction for vendor B. The percentage x remains 2.57 percent because it is the existing inventory carrying cost expressed as lines of work. We need the existing inventory carrying cost to calculate the ideal review frequency for vendors who do not offer volume discounts. Let's look at a comparison of the small wholesaler's total costs under the three considerations.

Look at figure 6.16. In case I we see that the small wholesaler's current review frequency costs 600 lines of work, 300 lines of work as an ordering cost plus 2.57 percent of the average inventory investment of $11,667. In case II we have applied the economic review frequency formula without regard to freight restrictions, and the total cost is 334 lines of work, 167 lines of work ordering cost plus 2.57 percent of an average $8333 inventory investment. In case III we see the application of the economic review frequency formula with a consideration for the $10,000 freight restriction on vendor A. In case III the total cost is 375 lines of work, 135 lines of work ordering cost and 2.57 percent of an average $9320 inventory investment.

Figure 6.16   A SMALL WHOLESALER: COST IN LINES OF WORK

		Ordering cost	Inventory carrying cost	Total cost
CASE I	Current review frequency	300	300	600
CASE II	Economic review frequency without freight restrictions	167	167	334
CASE III	Economic review frequency with freight restrictions	135	240	375

With the freight restriction there are 41 more lines of work for vendor A than there were when without the freight restriction. This means we paid 41 lines of work in order to meet the vendor's $10,000 minimum freight restriction. Even so we have saved 165 lines of work ordering cost and 60 lines of work inventory carrying cost. We have earned the free freight without adding to our existing 600 lines of work.

SUMMARY

The inventory manager uses the economic review frequency formula to reduce the ordering cost, to reduce the inventory carrying cost, and to meet freight restrictions.

At this point the inventory manager can either accept these review frequencies or choose from three other options. He can trade lines of work ordering cost for lower inventory carrying cost. He can trade lines of work inventory carrying cost for lower lines of work ordering cost. Or he can spend the potential savings for additional safety stock and improved customer service.

This economic review frequency formula is the only one I know of that combines ordering costs, inventory carrying costs, freight restrictions, volume discounts, and customer service. It comes up with a total cost reduction without estimates of ordering costs or inventory carrying costs. It is truly a new concept of economic order quantities.

The inventory manager uses the economic review frequency to determine the review time for each supplier. The focus forecasting inventory management system then adds this review time to the safety stock time and the lead time to get the reorder point time. Then it forecasts the reorder point quantity for the item over the reorder point time and decides how much to buy by subtracting the available quantity from the reorder point quantity.

This chapter completes the fourth step in the focus forecasting inventory management system. The controller has budgeted the total sales. The inventory manager has used the controller's sales budget to plan total inventory levels. The inventory manager, at the beginning of the month, has set the total safety stock levels to generate the purchases required to meet those total inventory levels. And now the focus forecasting inventory management system has used its logical ordering strategies and the new concept of EOQ to suggest buys for every item in the company's line.

We have already seen that focus forecasting calculates the vendor lead time by taking an average of the past three experiences with the vendor. We have seen how the inventory manager controls safety stock time, and now we see how the inventory manager uses this new concept of EOQ to control review time. Now let's see how the inventory manager controls lead time.

Chapter 7

Distribution Requirements Planning

A distribution requirements plan is a schedule of existing orders and future orders. The inventory manager sends this schedule to major suppliers, and the suppliers use it to plan production capacity and raw material availability. For the inventory manager the distribution requirements plan guarantees delivery and consistent lead time. For the supplier the distribution requirements plan eliminates forecasting and allows more time for production. This simple scheduling tool significantly improves inventory turnover and customer service for both the inventory manager and his supplier.

Material requirements planning has revolutionized manufacturing inventory control. Manufacturers used to forecast production requirements and raw material usage based on previous history. They treated requirements for production and raw material as though they were independent inventory control items. It didn't make sense because production requirements and raw material requirements depend completely on planned orders for finished goods. Material requirements planning generates production requirements and raw material requirements based on planned orders for finished goods. Today most manufacturers still forecast finished goods requirements as though they were independent inventory control items, but finished goods requirements clearly depend on major customers' planned orders.

Distribution requirements planning gives a schedule of existing orders and future orders to manufacturers. It is a continuation of the revolution of material requirements planning in inventory control.

Our company first started using the focus forecasting inventory management system in December 1972. If you remember, 1973 was a very difficult year for inventory control systems. It was hard for me to tell how much of our difficulty was due to the energy crisis and how much was due to our new inventory management system. I desperately needed someone to check the logic of our new inventory management system. So I turned to Ollie Wight; I've known him for many years, and I have tremendous respect for his logical mind.

Ollie listened intently as I explained focus forecasting. He had me repeat the system for his partner Walt Goddard. "There's one thing missing," Ollie said, "distribution requirements planning. Why surprise the vendor with an order once a month or once every week? You really don't have an answer for vendor lead time inconsistencies. Why not tell the vendor what you plan to buy over the next six months? Sure, you'll have to adjust your plan, but it has to be more accurate than a vendor's guess about what you are going to buy. If a vendor can put your planned purchases into his inventory plans he will give you better service. The vendor will give you better service and you won't have to fight his lead time inconsistencies." And so Ollie Wight gave me the idea of distribution requirements planning.

For a supplier distribution requirements planning is a godsend. It eliminates the need to forecast and the traditional reorder point inventory control system. The supplier's finished goods requirements planning becomes a simple extension of his material requirements planning system. For an inventory manager distribution requirements planning is the only way to control the delivery of goods from major suppliers. Distribution requirements planning is the only way the inventory manager can control lead time. Distribution requirements planning eliminates delivery problems before they happen. In this chapter you will see how the inventory manager sets up a distribution requirements plan to control lead time. In the next chapter you will see how the expediters use distribution requirements planning and other tools to insure the delivery of needed goods.

THE CONCEPT

A distribution requirements plan is a schedule of delivery dates and quantities for existing orders and future orders. The existing orders are the ones the inventory manager has already issued to suppliers. The future orders are those the inventory manager plans to issue. Existing orders are firm purchase order commitments; future orders are plans of future requirements.

The inventory manager sets up a distribution requirements plan with *major* suppliers for three reasons:

1. The top 20 percent of the suppliers in a company generally account for 80 percent of the purchase order volume. Any improvement in inventory turnover and customer service with these major suppliers means an overall improvement in inventory turnover and customer service for the entire company.
2. The major suppliers are the best equipped to use a distribution requirements plan. Distribution requirements planning is an extension of a manufacturer's material requirements planning system. Major suppliers with material requirements planning systems can feed a distribution requirements plan directly into their production and raw material schedules.
3. A distribution requirements plan is most effective when the company's purchases are a significant percentage of the supplier's total demand for an item. Since a company's total purchases are concentrated in major suppliers, its purchases should be a significant percentage of a supplier's demand for an item. The larger the company's purchases are of its supplier's total volume, the more valuable the distribution requirements plan will be.

Setting up a distribution requirements plan is a negotiating process. My company's experience has shown that a distribution requirements planning system works only when the inventory manager and the supplier have agreed on four points:

1. The inventory manager agrees to make firm purchase order commitments far enough in advance that the supplier can guarantee delivery.
2. The inventory manager agrees to send the supplier a schedule of future orders every time he reviews the supplier for a buying decision.
3. The supplier agrees to ship all purchase orders on the distribution requirements planning system complete and on the scheduled delivery date.
4. The supplier agrees to ship any late orders as soon as they are available and to pay the freight for shipping them.

These four points put teeth into the distribution requirements planning system.

In business there is no contract until money exchanges hands. That's why you always see in contracts "for due consideration." The inventory manager could send a schedule of existing orders and future orders to any of the major suppliers, but a distribution requirements planning system exists only after the inventory manager and the supplier have agreed on all four points.

This agreement provides major advantages to the inventory manager and to the supplier. The inventory manager invests in safety stock for two reasons: to cover low forecasting errors and to cover late supplier deliveries. This agreement lets the inventory manager either eliminate the safety stock investment used to cover late supplier deliveries or use it to improve overall customer service.

The supplier invests in safety stock for two reasons: to cover delays in the manufacturing process and to cover his own low forecasting errors. This agreement lets the supplier either eliminate the safety stock used to cover low forecasting errors or use it to improve customer service.

A distribution requirements plan with these four points of agreement improves inventory turnover and customer service for both the inventory manager and the supplier. With distribution requirements planning the inventory manager is solely responsible for forecasting, and the supplier is solely responsible for consistent lead time. This is the way it should be.

DISTRIBUTION MECHANICS

In the earlier chapters we saw how the focus forecasting inventory management system projected demand and suggested buys for a broiler pan. Now we return to this example to see how the distribution requirements planning system works. First we review the buying, shown in figure 7.1.

Figure 7.1   HOW TO BUY A BROILER PAN/JULY 1 THROUGH DECEMBER 1
THIS YEAR

Review time	1.0 months				
Lead time	1.5 months				
Safety stock time	1.2 months				
Reorder point time	3.7 months				

	July	August	September	October	November	December
Current on hand	200	179	206	206	206	206
Current on order	100	148	121	121	121	121
Forecasted demand	121	121	121	121	121	121
Reorder point	448	448	448	448	448	448
Current available	300	327	327	327	327	327
Purchase order	148	121	121	121	121	121

Let's again go through the exercise of calculating the average broiler pan inventory investment. On September 1 we have 206 pans. We sell 61 pans between September 1 and September 15. No pans arrive between September 1 and September 15. So the low point is 145 pans on September 15. On September 15 we had 145 pans. On September 16 the 121 pans we ordered on August 1 arrive. So the high point is 266 pans on September 16.

During the month the inventory ranges from a low of 145 broiler pans to a high of 266 broiler pans. The average inventory investment then is 206 broiler pans, or an average of 1.7 months on hand. This 1.7 months on hand is one-half the review time (1 month) plus all the safety stock time (1.2 months). Thus the average inventory investment is 1.7 months. The average inventory investment is always one-half the review time plus all the safety stock time. The length of the vendor's lead time does not affect the inventory investment.

Inconsistencies in the vendor's lead time, however, do affect inventory investment. If the goods arrive from the vendor one-half month early, we will have one-half month's additional inventory investment. If the goods arrive from the vendor one month late we could very well have run out of inventory. Under this buying approach it is very unlikely that the lead time will be precisely 1.5 months. The 1.5 month lead time used in this reorder point is simply an average of the past experiences on delivery of all items from this vendor.

The inventory manager decides to put the broiler pan vendor on a distribution requirements planning system. The inventory manager agrees to make firm purchase order commitments 2.5 months into the future. This gives the vendor an additional month to deliver the needed goods. The inventory manager agrees to give the broiler pan vendor a projection of requirements six months into the future. This lets the broiler pan vendor plan production and raw material requirements for the future. The broiler pan vendor guarantees delivery on a specific date and agrees to ship late orders as soon as they are available. The broiler pan vendor agrees to pay the freight involved in shipping late orders.

Let's see the impact on inventory investment using a distribution requirements planning system for buying a broiler pan.

Look at figure 7.2. On July 1 the inventory manager makes the regular purchase using the existing vendor lead time; then changes the lead time to 2.5 months. The inventory manager uses the focus forecasting inventory management system to generate another purchase order on July 1 with a scheduled delivery of September 15. That's why figure 7.2 shows two columns for July. This is the same purchase order that would have been generated on August 1 for delivery on September 15. Since the 121 broiler pans will arrive on September 15 there is no change in the broiler pan inventory investment.

In each month after that the inventory manager generates a

Figure 7.2 HOW TO BUY A BROILER PAN/DISTRIBUTION REQUIREMENTS
PLANNING/JULY 1 THROUGH DECEMBER 1 THIS YEAR

	July	July	August	September	October	November	December
Review time	1.0	1.0	1.0	1.0	1.0	1.0	1.0
Lead time	1.5	2.5	2.5	2.5	2.5	2.5	2.5
Safety stock time	1.2	1.2	1.2	1.2	1.2	1.2	1.2
Reorder point time	3.7	4.7	4.7	4.7	4.7	4.7	4.7
Current on hand	200	200	179	206	206	206	206
Current on order	100	248	269	242	242	242	242
Fore-casted demand	121	121	121	121	121	121	121
Reorder point	448	569	569	569	569	569	569
Current available	300	448	448	448	448	448	448
Purchase order	148	121					
Planned require-ments			121	121	121	121	121
Sched-uled delivery	8/15	9/15	10/15	11/15	12/15	1/15	2/15

purchase order using the vendor's promised lead time of 2.5 months. The purchase orders under the distribution requirements planning system are the same as the purchase orders under the normal logical ordering strategies. The only difference is that the purchase orders are generated one month earlier.

Notice that the current on hand remains the same under both systems. The only thing that changes is the current on order. The inventory investment remains one-half the review time plus the safety stock time. Now, however, the safety stock time no longer has to cover inconsistencies in vendor lead time. At this point the inventory manager can either reduce the safety stock time on broiler pans or use the same safety stock time to improve customer service on broiler pans.

Distribution requirements planning has not changed the focus forecasting logical ordering strategies. The reorder point quantity is still the forecasted demand over the reorder point time. The reorder point time is still the review time plus the safety stock time plus the lead time. The only thing that has changed is that the lead time is now a promised 2.5 months rather than an average 1.5 months.

The inventory manager runs two purchase orders on July 1 to start up the distribution requirements planning system. These two purchase orders are for all the items that the inventory manager buys from the broiler pan vendor. The inventory manager mails these two purchase orders along with the first distribution requirements plan to the broiler pan vendor on July 1. Let's look at the distribution requirements plan now.

The distribution requirements plan in figure 7.3 shows that the vendor has promised a 2.5 month guaranteed delivery. If a purchase order meets a dollar amount of $10,000 the broiler pan vendor ships freight prepaid. From September 15 on, the broiler pan vendor must ship backorders on any item immediately at his expense. The distribution requirements plan shows the purchase order date and purchase order number on all firm purchase orders and the planned order date on all planned future orders. It shows the scheduled delivery for all orders and every item that the inventory manager orders from the broiler pan vendor.

Notice that the total dollars on each order do exceed the freight prepaid minimum of $10,000. The inventory manager has selected a review time for the broiler pan vendor that meets the freight prepaid restriction. Let's look at each column on the distribution requirements plan.

Column 1 shows late orders. Starting with September 15 late orders are any orders more than 2.5 months old. All other orders will be listed in detail on the distribution requirements plan.

Column 2 shows the July 1 purchase order that the inventory manager issued under the original 1.5 month lead time.

Column 3 shows the July 1 order issued under the distribution requirements planning lead time of 2.5 months.

Columns 4–8 show the planned purchase orders from August 1 through December 1. On August 1 the inventory manager will issue a firm purchase order for delivery on October 15. On August 1 the inventory manager will run a new distribution requirements plan revising the quantities and scheduled delivery dates for the remainder of the future orders. Each time the inventory manager reviews this vendor's items for a buying decision he sends the vendor a new firm purchase order and a new distribution requirements plan.

Figure 7.3 DISTRIBUTION REQUIREMENTS PLAN FOR A BROILER PAN VENDOR

| July 1 2.5 month guaranteed delivery | | | | | | $10,000 freight prepaid | | |
	1	2	3	4	5	6	7	8
Purchase order date	6/1	7/1	7/1					
Planned order date	27461R1	27462R1	27463R1	8/1	9/1	10/1	11/1	12/1
Scheduled delivery	7/14	8/15	9/15	10/15	11/15	12/15	1/15	2/15
Item description				*Order quantities*				
Broiler pans 10072	100	148	121	121	121	121	121	121
Cookie sheets 10073	0	0	52	85	85	85	85	85
				Other items				
Total dollars	$683	$10,252	$10,282	$10,126	$10,142	$10,093	$10,492	$10,112

Attention: *A Broiler Pan Vendor*
Ship purchase orders with a scheduled delivery prior to 7/1
  immediately freight prepaid

Now let's look at the way the broiler pan vendor uses the distribution requirements plan in his material requirements planning system.

SOME MECHANICS OF MATERIAL REQUIREMENTS PLANNING

Today manufacturers are using some very sophisticated material requirements planning systems. This example will give you a general idea of how a manufacturer uses a distribution requirements plan in conjunction with a material requirements planning system.

Look at figure 7.4. The vendor has calculated an economic order quantity of 500 units for a broiler pan and the manufacturing lead time is 2.0 months. The demand requirements are directly from the inventory manager's distribution requirements plan. The on-hand inventory shows the vendor's inventory balance at the beginning of

Figure 7.4 MATERIAL REQUIREMENTS PLAN: A BROILER PAN

July 8
Economic order quantity: 500
Manufacturing lead time: 2.0 months

	Actual demand			Planned demand				
	1	2	3	4	5	6	7	8
	Late	8/15	9/15	10/15	11/15	12/15	1/15	2/15
Demand requirements	0	148	121	121	121	121	121	121
On-hand inventory	400	252	131	10	389	268	147	26
Production orders	0	0	500	0	0	0	0	0
Scheduled delivery	—	—	11/15	—	—	—	—	—

each period. The vendor schedules production orders based on the on-hand inventory balance needed to service demand requirements. Let's look at this material requirements plan column by column.

Column 1: The July 8 demand requirements are zero. The inventory manager's distribution requirements plan showed 100 broiler pans were still owed. This means the broiler pan vendor shipped the 100 broiler pans since July 1. The July 8 inventory balance is 400 units. There are no orders put into production on July 8.

Column 2: The August 15 demand requirements are 148 units. The August 15 on-hand inventory balance of 252 units is the July 8 on-hand inventory balance of 400 units less the August 15 demand requirements of 148 units. No orders are put into production on August 15.

Column 3: The September 15 demand requirements are 121 units. The September 15 on-hand inventory balance of 131 units is the August 15 on-hand inventory balance of 252 units less the September 15 demand requirement of 121 units. On September 15 the broiler pan manufacturer schedules 500 units into production for delivery on November 15. Notice that by October 15 the on-hand inventory balance is down to 10 units, so the vendor needs these 500 units to service the November 15 demand requirement.

Column 4: The October 15 demand requirements are 121 units. The October 15 on-hand inventory balance of 10 units is the September 15 on-hand inventory balance of 131 units less the October 15 demand requirement of 121 units. No orders are put into production on October 15.

Column 5: The November 15 demand requirements are 121 units. The November 15 on-hand inventory balance of 389 units is the October 15 on-hand inventory balance of 10 units plus the September 15 production order of 500 units less the November 15 demand requirement of 121 units. No orders are put into production on November 15.

Columns 6–8: The December 15, January 15, and February 15 on-hand inventory balances show the effect of subtracting the demand requirements from the November 15 on-hand inventory balance of 389 units.

When the vendor gets the inventory manager's distribution requirements plan for August he may move up the production of the 500 units scheduled for delivery on November 15. If the inventory manager's next distribution requirements plan increases the October 15 demand requirement more than 10 units, the vendor would run out of stock unless he moved up the scheduled delivery of the September 15 production order.

With a 2.0 month manufacturing lead time the vendor does not have to make this decision at this time. In fact, the inventory manager's next distribution requirements plan may show a marked decrease in the October 15 demand requirements. The vendor may then delay the scheduled delivery of the September 15 production order.

This is how the vendor uses the inventory manager's distribution requirements plan in his material requirements planning system. Let's go one step further now and see how the vendor's production orders affect his raw material planning. Let's look at one of the raw materials that go into a broiler pan.

Each broiler pan takes 2.2 broiler pan handles on the average.

Look at figure 7.5. The vendor reviews the handle inventory once a month, so the review time is 1.0 months. Lead time on broiler pan handles is 1.0 months; it takes one month to receive broiler pan handles in stock once a buying decision is made.

The material requirements are the broiler pan production orders converted into broiler pan handles. The on-hand inventory shows the number of broiler pan handles in stock at the beginning of each period. The purchase orders are shown along with their scheduled delivery dates. Let's look at each column in the material requirements plan.

Column 1: The July 15 material requirements are zero, and the July 15 on-hand inventory balance is 200 units. On July 15 the raw material buyer places a purchase order for 900 units scheduled for

Figure 7.5 MATERIAL REQUIREMENTS PLAN: BROILER PAN HANDLES

July 15
Review time: 1.0 months
Lead time: 1.0 months

	1	2	3	4	5	6	7	8
	Late	8/15	9/15	10/15	11/15	12/15	1/15	2/15
Material requirements	0	0	1100	0	0	0	0	0
On-hand inventory	200	1100	0	0	0	0	0	0
Purchase orders	900	0	0	0	0	0	0	0
Scheduled delivery	8/15	0	0	0	0	0	0	0

August 15 delivery. The raw material buyer sees that there are no material requirements until September 15. When the raw material buyer places the purchase order for scheduled delivery on August 15 he is building one month's safety stock. This safety stock will be required if the broiler pan vendor moves up the planned production order from September 15 to August 15.

Column 2: The August 15 material requirements are zero. The August 15 on-hand inventory balance of 1100 units is the July 15 on-hand inventory balance of 200 units plus the July 15 purchase order of 900 units. No purchase orders are placed on August 15.

Column 3: The September 15 material requirements are 1100 units. The September 15 on-hand inventory balance is zero: the August 15 on-hand inventory balance of 1100 units less the September 15 material requirements of 1100 units. No purchase orders are placed on September 15.

Columns 4–8: No raw material activity is planned from October 15 through February 15 at this time.

The inventory manager may increase the distribution requirements plan on August 1, and this increase may cause the broiler pan vendor to place a production order some time in January or February. A planned production order for January or February would project material requirements for broiler pans some time in January or February. Then the material requirements plan for broiler pan handles would alert the raw material buyer to buy additional broiler pan handles some time in the future.

If the broiler pan vendor buys handles from a major supplier he may well set up a distribution requirements plan for handles with that supplier, and the material requirements plan for handles would become the distribution requirements plan with the handle supplier.

When you think about it the only independent demand occurs at the consumer level, when someone goes to the retail store and decides to buy something. All other demand is really a summary of a customer's future requirements. The retailer could ask all the individuals in the area what their future requirements would be. Then the retailer wouldn't have to forecast future demand—he could just add up the future requirements. The wholesaler could ask all his retail customers what their future requirements would be, so the wholesaler wouldn't have to forecast future demand. The manufacturer could ask his manufacturing customers, wholesale customers, and retail customers what they thought their future demand would be. Then the manufacturer wouldn't have to forecast future demand. Whenever the benefits justify the expense, distribution requirements planning should replace forecasting of future requirements.

This example shows how the inventory manager's distribution requirements plan is used in the vendor's material requirements planning system. Distribution requirements planning improves turnover and customer service both for the inventory manager and for the vendor. Materials requirements planning has revolutionized manufacturers' inventory control, and distribution requirements planning will continue this revolution.

SUMMARY

A distribution requirements plan is a schedule of future orders and their required delivery dates. The schedule includes firm orders that the inventory manager has already issued and planned orders that the inventory manager will issue in the future. The inventory manager sends the distribution requirements plan to the supplier, who uses it with his material requirements planning system to insure production capacity and raw materials availability.

Distribution requirements planning improves turnover and customer service for the inventory manager. It eliminates the need for safety stock to cover inconsistencies in lead time, so the inventory manager can either eliminate this safety stock or use it to improve customer service.

Distribution requirements planning also improves inventory turnover and customer service for the supplier. Part of the supplier's safety stock inventory investment is used to cover low forecast errors. Since the inventory manager gives the supplier a firm commitment for future purchases, the supplier no longer needs safety stock to cover these errors. The supplier can either eliminate this safety stock or use the safety stock to cover delays in the manufacturing process.

Distribution requirements planning works only when the inventory manager and the supplier have a firm agreement that includes these four points:

1. The inventory manager agrees to make firm purchase order commitments far enough in advance that the supplier can guarantee delivery.
2. The inventory manager agrees to send the supplier a schedule of future orders every time he reviews the supplier for a buying decision.
3. The supplier agrees to ship all purchase orders on the distribution requirements planning system complete and on the scheduled delivery date.
4. The supplier agrees to ship late orders as soon as they are available and to pay the freight involved in shipping them.

Distribution requirements planning makes the inventory manager solely responsible for forecasting future requirements and makes the supplier solely responsible for consistent lead time. This is the way it should be.

In this chapter we have seen how distribution requirements planning lets the inventory manager control lead time. In the next chapter we will see how the distribution requirements planning system eliminates delivery problems before they happen. We will see how the expediters solve the delivery problems that do happen, insure the delivery of needed goods, and see that delivered goods move quickly from the receiving dock to the finished goods shelves.

Chapter 8

How to Use Expediting Effectively

SPEEDING UP DELIVERY

The expediter's job is to insure the delivery of needed goods. Theoretically the expediter's job shouldn't exist. Suppliers want to ship goods, because their whole business depends on sales. Shippers want to move goods, because their whole business depends on their ability to move goods quickly. Receiving departments want to receive goods and put them away as quickly as possible, because the fewer goods they have on the dock, the easier their job is. So a company shouldn't need any expediters.

In the real world, though, things don't go according to plan; anything that can happen will happen. The inventory manager's forecast could be low. Sometimes he needs the goods earlier than planned. Sometimes the supplier doesn't have enough goods to service the customers' orders. Sometimes the supplier even loses customers' orders. Freight companies consolidate orders to fill their trucks and their freight cars. Sometimes they lose customers' orders. The receiving department is working against budgets. Sometimes there just are not enough people to handle all the actual goods they receive. So in the real world there must be expediters to insure the delivery of needed goods.

So far we have seen the first four steps of the focus forecasting inventory management system. The controller budgeted the total sales. The inventory manager used the controller's sales budget and historic turnover service levels to plan total inventory levels for the future. The inventory manager set total safety stock levels to generate purchases to meet those total inventory levels. The buyers used the focus forecasting inventory management system's suggested buys to purchase all the items in the company's line. The inventory manager has already taken steps to insure the delivery of needed goods by setting up, wherever possible, a distribution requirements plan with major suppliers. Now it is up to the expediters to insure the delivery of needed goods from the suppliers. It is up to the expediters to insure that goods arriving at the receiving dock are put away quickly.

This chapter is about a *professional* expediting staff. One professional expediter is worth four clerical people recording vendor promises. During the energy crisis the *Wall Street Journal* ran an article about one such professional expediter. "George Galla, Corporate Expediter, is busy coaxing, pushing suppliers. Mr. Galla is an expediter, one of the business world's hottest brands of specialists in these days of great demand and little supply. As such he uses a mixture of cajolery, diplomacy, personal friendships, and knowledge of his industry to see that his company secures the goods it needs from its suppliers."

Professional expediters need to know which orders are needed first and where the orders are. This chapter describes an order control system that provides this information. In this system the computer logs each purchase order and for most it calculates a priority date. As time passes the computer updates the priority dates based on actual customer demand. These priority dates tell the expediters, the suppliers, and the operations department which orders are needed first. This system also lets the operations department record the arrival of needed goods and their movement from the receiving dock to the finished goods shelves. It produces expediting reports, which tell the expediters and the suppliers which orders have not yet arrived, and float reports, which tell the expediters and the operations department which orders have not yet been put away. Together these reports tell the expediters where the orders are.

This expediting system is not unique. But this chapter should give you some insight into some of the things people can do to make an expediting system successful and how you can use expediting effectively.

THE PROFESSIONAL EXPEDITER

"Hello, Harry, this is Dick VanPatten from Giant. Boy, my boss has really been on my back about those mowers. No kiddin'. I'm in deep trouble. I know you're trying for me, but if I don't get those mowers before the end of June my goose is cooked. How about shipping part of the order and getting me off the hook? Hey, did I tell you that Margaret and I just had another baby, Thanks. Yeh, that makes five. Hey, yeh, that would be great. If you could ship the stock part of the order, at least we'll have some mowers in the barn. See ya now, Harry. Thanks again."

"Hello, Weeter Valve? Mr. Richard VanPatten calling for your Mr. Weeter. Mr. Weeter? Hold just one moment for Mr. Richard VanPatten of Giant Hardware, Inc."

"VanPatten here, Mr. Weeter. I don't like to call you direct, but you know your private label valves are the backbone of our valve business. Your people don't seem to understand. What? A raw material shortage? Hmmm, Value Line Valve saw Mr. Burton the other day. They seem to think they can ship all the valves we'd ever use. No, no, I don't think we'd ever switch. Weeter Valves are quality, but you have our backs to the wall. We're between a rock and a hard place. You and I both know we're almost 10 percent of your valve business. Yes, we've known each other a long time. Now take care of us. We need those orders you've been holding. Order numbers? No, please don't trouble yourself with the details. Your Mr. Clark knows the order numbers. Fine. I'll wait for his call this afternoon."

"Give me Chuck Hague. Chuckie? Hey listen, Chuckie, you told me those flagpoles were on the way. What gives? What do you mean, who's this? Geez, do you tell everybody they're on the way? C'mon, no more friggin' around. Our wagon will be in there at 5:00 this afternoon. Load it up with whatever you got. What? You still don't know? It's Dickie VanPatten from Giant, ya dummy. No I ain't got no hangover. The only pain I got is you and it ain't in my head. Yeh, 5:00 today, right. That's right. Yeh, I'll see ya at the Roadster Outing. Now don't forget, 5:00. Load it up, baby. Load it up."

Expediting costs money. There's a real temptation to ignore the people involved and to cut the expediting short. You might think you should just give the order numbers and get the shipping dates. But expediting means getting goods faster; it doesn't mean writing down what's happening. Expediting is people influencing other people to ship your company's goods first. Expediting is getting someone—a Harry, a Mr. Weeter, a Chuckie Hague—to put your company's order on top of the pile and to leave it there when somebody else calls later.

The same expediter should always talk to the same suppliers. The expediter should learn about the supplier's company, products, sales volume, shipping procedures, location, prices, selling plans, executive structure, number of employees, credit rating, and anything else that helps the expediter understand the supplier. More, the expediter should learn about the supplier's customer service representative—age, education, family, hobbies, length of service, attitude, politics, and anything else that helps the expediter understand the customer service rep.

An expediter's only consistent power is his rapport with the suppliers' customer service rep. Once in a while, the expediter can put pressure on a supplier by calling the supplier's sales manager or by calling the supplier's company president. But expediting is an everyday job. The expediter must use the supplier's customer service rep because he's the one who can put your order on top of the pile.

It's easy for an expediter to find out about a supplier's company. There's Dun & Bradstreet, annual reports, the supplier's sales representative, and the supplier's advertising material. But how does an expediter get to know the supplier's customer service representative over a telephone? These two people will probably never meet face to face. The expediter who just calls up and writes down order numbers, shipping dates, pro numbers, and freight lines doesn't get to know the customer service representative as a person. This expediter is just a scorekeeper. A company would be better off without this expediter; the company would save money.

It boils down to taking the time to treat the supplier's representative as a person. The call should start like this: "Hello Harry, this is Dick VanPatten." The call should not start like this: "Hello Massive Mowers. This is Giant Hardware." People make decisions to move goods. A person puts your order on the top of the pile. Expediting is a personal thing.

In this chapter we will see an expediting system that establishes priority dates and tells the expediter which goods to get first. But without a personal relationship the expediting system is just so much window dressing. An expediter's only consistent power is his rapport with the supplier's customer service rep. This personal relationship is the difference between an expediting clerk and a professional expediter.

AN EXPEDITING SYSTEM

The expediting system is a communications system with simple goals. It answers two questions: (1) What are the orders we need now? (2) Where are those orders now?

The expediting system uses priority dates to tell us which orders we need now. Some orders have a fixed priority date. In the distribution requirements planning system the fixed priority date is the vendor's promised delivery date. The fixed priority date for orders for promotions and other special events is the date the goods must arrive to service the special event. The expediting system calculates a priority date on all other orders based on maintaining a given level of customer service.

The expediting system uses a cathode ray tube system to tell us where the orders are now. In every receiving location is a cathode ray tube. When an order arrives the receiving clerk keys in the purchase order number to generate receiving papers, which are used to receive the vendor's order. At the moment the receiving clerk keys in the purchase order number, the expediting system records the arrival of the goods. A cathode ray tube in the expediting department lets the expediter find out that a purchase order has arrived as soon as the receiving clerk keys in the purchase order number.

The expediting system uses priority dates to tell us which orders we need now. The priority date tells the expediters which orders to expedite first and tells the receiving department which orders to put away first. There are two kinds of priority dates, fixed priority date and calculated priority date. The delivery date that the vendor has promised is a fixed priority date. When the inventory manager buys goods for a promotion or other special event, the date by which those goods must arrive on the shelf is a fixed priority date. A calculated priority date is the date by which the goods must arrive on the shelf to maintain a given customer service level for all the items on that vendor's purchase order. Both fixed and calculated priority dates tell us which orders we need now.

It's easy to see how the inventory manager sets a fixed priority date for promotion orders and special events. The next section explains how the expediting system calculates a priority date for other orders.

PRIORITY DATE CALCULATIONS

The calculated priority date is the date by which the vendor's order must arrive on the shelf to meet the desired line-fill rate. The inventory manager evaluates customer service in terms of line-fill rate. You will recall from chapter 4 that the line-fill rate is the number of lines the inventory manager is able to ship divided by the number of lines that customers ordered.

Suppose the inventory manager's line-fill rate goal is 90 percent. Then the calculated priority date on any purchase order is the date by which that purchase order must arrive to meet that goal for all the items on that purchase order. To see how this works, let's look at an expediting request for a particular purchase order, shown in figure 8.1.

The expediting system prints an expediting request so the expediting department knows which orders are needed now. This expediting request shows a purchase order issued to Weeter Valve from Giant Hardware on May 1. Today is July 1 and the order still has not arrived. The priority date on this order is July 15. That

Figure 8.1 EXPEDITING REQUEST

		7/1/this year	
To:	Mr. Robert Clark	Telephone 617-555-1518	
	Weeter Valve		
	310 Dayton Street		
	Cleveland, OH 78003		
From:	Richard VanPatten	Telephone 412-555-6787	
	Giant Hardware		

Please ship the goods you owe us at once.

Vendor number	Our order number	Our order date	Priority date
2698E	7098	5/1/this year	7/15/this year

Item number	Item description	On-order quantity	Each price	Extended amount
7438	1 × ¾" valve	3600	0.50	$1800
7439	2 × ¾" valve	4200	0.60	2520
7440	3 × ¾" valve	3600	0.75	2700

Notes: 6/15 Mr. Clark promised delivery by 6/22
7/1 Called Mr. Weeter

means that if this order does not arrive by July 15 the line-fill rate on the three items on this order will drop below 90 percent. Let's see how the expediting system calculated the July 15 priority date on this order.

First, remember that the inventory manager's goals are in terms of line-fill rate, but all the information in the focus forecasting inventory management system is in terms of units of items. So we have to convert units to lines. The focus forecasting inventory management system keeps track of the average number of units a customer orders each time he orders an item. Let's suppose that a customer orders 10 units at a time of the items on the Weeter Valve order. Thus if the inventory manager has 2200 units on hand for item 7438, he has 220 lines on hand. If the forecasted demand per week for item number 7438 is 600 units per week, the demand is 60 lines per week.

The expediting system uses the lines on hand for each item and the forecasted line demand by week for each item to calculate the priority date for all the items on the Weeter Valve order. Let's look at this priority date calculation shown in figure 8.2.

Figure 8.2 WEETER VALVE ORDER/PRIORITY DATE CALCULATION/LINE-FILL RATE GOAL 90%

7/1/this year

Step 1

Item number	Lines on hand	Forecasted line demand by week							
		7/8	7/15	7/22	7/29	8/5	8/12	8/19	8/26
7438	220	60	60	60	60	60	60	60	60
7439	180	80	80	80	80	80	80	80	80
7440	300	60	60	60	60	60	60	60	60
Total line demand		200	200	200	200	200	200	200	200

Step 2

Item number	Lines on hand	Forecasted line cancellations by week							
		7/8	7/15	7/22	7/29	8/5	8/12	8/19	8/26
7438	220	0	0	0	20	60	60	60	60
7439	180	0	0	60	80	80	80	80	80
7440	300	0	0	0	0	0	60	60	60
Total line cancellations		0	0	60	100	140	200	200	200

Step 3

	Line-fill rate by week (%)							
	7/8	7/15	7/22	7/29	8/5	8/12	8/19	8/26
All items	100	100	70	50	30	0	0	0

The priority date is the week before the line-fill rate drops below 90%. The priority date is 7/15/this year.

Step 1 shows all three items on the Weeter Valve order, their lines on hand, and their forecasted line demand by week. The expediting system totals the line demand for each of these items to get the total line demand by week for all the items on the Weeter Valve order.

Step 2 shows the forecasted line cancellations by week. Let's look at one of the items on the Weeter Valve order to see how the expediting system calculated the forecasted line cancellations by week.

Item 7438 has 220 lines on hand on July 1. The forecasted line demand for July 8 is 60 units. At the end of the week of July 8 there will be 160 lines on hand. The forecasted line demand for the week of July 15 is 60 lines, so on July 15 there will be 100 lines on hand. The forecasted line demand for the week of July 22 is 60 lines. At the end of the week of July 22 there will be 40 lines on hand. The forecasted line demand for the week of July 29 is 60 units. This means that the inventory manager can ship only 40 of the lines demanded for the week of July 29 for item 7438. The forecasted line cancellations for the week of July 29 are then 20 lines.

The expediting system calculates the forecasted line cancellations by week for each item on the Weeter Valve order, and totals the line cancellations for all the items on the Weeter Valve order.

In step 3 you see the line-fill rate by week for all the items on the Weeter Valve order. In the week of July 8 the inventory manager will be able to ship all the lines that customers order. So the line-fill rate is 100 percent. In the week of July 15 the inventory manager will still be able to ship all the lines that customers order, so the line-fill rate remains 100 percent. In the week of July 22 the inventory manager will be able to ship only 140 of the 200 lines that customers order, so the line-fill rate will drop to 70 percent. The priority date is the week before the line-fill rate drops below 90 percent, in this example July 15 of this year.

The distribution requirements planning system puts a fixed priority date on every order. The inventory manager puts a fixed priority date on every order issued for a promotion or special event. The expediting system calculates a priority date for every order not on the distribution requirements planning system and for every order not needed for a promotion or special event. The expediting system uses a priority date to tell us which orders we need now. The priority date tells the expediting department which orders to expedite and tells the receiving department which orders to put away first.

Let's now look at how the expediting system tells us where the orders are now.

THE FLOAT REPORT

"Hello, Mr. VanPatten? This is Mr. Weeter from Weeter Valve. You called me this morning about delivery. *Your* truck picked up *your* orders seven days ago. The valves are in *your* warehouse. Nice talking to you. Call me again if I can help you."

Say a company buys $50 million a year from its vendors. That's $1 million a week. If the company's receiving department gets behind it's a serious problem. One week behind is a million-dollar backlog, a million-dollar float. So it is important to know where orders are. How does the expediting system keep track of where the orders are now?

In each receiving location is a cathode ray tube, essentially a TV screen. Attached to the TV screen is a typewriter keyboard and a slow-speed printer. The TV screen, the typewriter keyboard, and the slow-speed printer are hooked by a telephone line to a central computer, which has a record of all open purchase orders. When a truck comes in with a shipment, the receiver unloads the truck, finds the purchase order number and the vendor number on the packing list, and types the purchase order number on the cathode ray tube. On the screen the computer shows the whole open order. The receiving clerk checks the items against the packing slip. If it's correct, he asks the computer to print a receiving document.

Now the expediting system knows where the order is. The expediting system also prints a priority date at the top of the receiving document which tells the receiving clerk that the Weeter Valve order, for example, must be put away by July 15. There may be 400 orders on the receiving department floor and another 100 orders may have just arrived, but every order has a priority date that determines which orders have to be put away first.

Each day the expediting system prints a report showing every order that has arrived and every order that is on the receiving department floor in priority sequence. Orders that have arrived but have not been put away are in float. The expediting system prints a float report showing every purchase order that has arrived but not been put away in priority date sequence.

Figure 8.3 is an example of a float report. There are 414 orders totalling $1,096,482 in float. The receiving department recorded that the Weeter Valve order arrived on July 2. The Weeter Valve order's priority date is July 15, but there are five other orders with earlier priority dates.

Figure 8.3   FLOAT REPORT

Priority date	Date arrived	Purchase order	Vendor number	Vendor name	7/17/this year Dollar amount
5/21/TY	7/1/TY	7349A	7597	Red Angel	$ 178.00
5/30/TY	7/5/TY	4264E	6408	Hooker Pin	4976.50
6/5/TY	7/12/TY	7069G	3902	United Nail	2014.80
6/12/TY	7/17/TY	1067C	7190	Stanley Screw	2834.90
6/28/TY	7/12/TY	3482G	6066	Leaf Raker	148.96
7/15/TY	7/2/TY	2698E	7098	Weeter valve	6020.00

Number of orders: 414
Total float: $1,096,482

The float report tells the receiving department which purchase orders to put away first. In addition, once the expediting system knows that a purchase order has been included in the float report, it no longer prints an expediting request for a vendor. If the Weeter Valve order had been recorded in float on July 1 it would not have appeared on the July 1 expediting request.

These priority dates tell the receiving department which sequence for putting away orders is best for customer service. Unfortunately, the receiving department may have priorities that override customer service priority dates. This is why the expediting department must not only see that the vendor gets the orders to the company, but also make sure that the receiving department puts the right orders away first. Let's look at some of the receiving department's priorities.

THE RECEIVING DEPARTMENT'S PRIORITIES

Every day the receiving department manager looks at the float report. The receiving department manager knows the importance of priority dates. He knows the customer comes first. When he sees warning signs on the float report, he calls in the receiving supervisor.

"Bud, look at this float report. Red Angel has been sitting here for two weeks. It's the number one priority in the whole receiving department. Hooker Pin has been laying here almost as long and

United Nail is still here after almost a week. What's the problem?"

"Yeh, Jim, I know, but we had to work on some other orders."

"What do you mean, you had to work on some other orders? Expediting is going to be all over us if you don't get those pins and nails put away."

"Well Jim, we had a whole truckload of Remington guns and arms sitting on the receiving dock. That stuff is top-security. Did you want to let it sit on the dock?"

"No I don't, but you've got twelve people in that receiving area. What are the rest of them doing?"

"Well, we had three cars of bicycles come in Friday. If we didn't get them unloaded the railroad would charge us demurrage. Three cars sitting here for a week would cost us over $1500."

"Yeh, Bud, but you've still got twelve people. And those pins and nails have been sitting here for two weeks."

"Did you see those spring promotion extension ladders, Jim? We had thirty-footers and forty-footers laying all over the dock. We can't carry the nails over the ladders. We had to move the ladders."

"OK, OK, Bud. Just tell me, when are you going to get those nails and pins put away?"

The receiving manager knows the customer comes first, but he has some overriding priorities. If he used priority dates in strict sequence he would destroy the receiving department efficiency. Eventually he would not be able to put anything away. He knows the importance of priority dates, but he must also consider efficiency. He must keep down demurrage costs, put away security merchandise, and must move bulk goods to get at the priority goods.

Even though the receiving department manager is under strong pressure to maintain efficiency, the expediting department must monitor the float report to make sure that the receiving department's need for efficiency doesn't destroy customer service.

The main expediting job is to get goods shipped first. The second is to make sure that the goods keep moving when they arrive at the receiving dock. The third is to get information about the movement of goods to the customer. When goods are scarce it becomes as important as the other two. There is a right way to communicate information from expediting to the customer and there is a wrong way. Many companies make the mistake of letting their expediting department become a customer service department.

EXPEDITING AND CUSTOMER SERVICE

Customer calls about delivery problems should *not* go directly to the expediters. Here's why.

"Hello, this is Harry Saul, customer 8743."

"Hello, Mr. Saul. How can we help you?"

"When will we get some antifreeze?"

"We won't have any stock in for the whole season. We are sold out."

"Wait a minute. Max's Hardware down the block from me just got half a truckful. He's a big customer, right? He gets antifreeze and I don't."

"Well sir, without checking into the details, I'd have to say that Max probably ordered at the spring market. That was six months ago. The vendors have guaranteed those spring market orders."

"It's like I said, he's a big customer and I'm not. Well, anyway, answer me this. My delivery is supposed to be on Wednesday morning. Here it is Thursday afternoon and no truck. I had high school kids here yesterday to unload that truck. Where is it?"

"Did you order antifreeze at the spring market, sir?"

"I'm not talking about antifreeze. I'm talking about my whole order."

"Oh, I'm sorry, sir, we only handle expediting. You have to talk to our trucking department to find out what happened to your regular delivery. Hold on a minute, I'll transfer your call."

Bzzz, bzzz, bzzz.

"Sir, that line is busy. I'll write it down and have them call you."

"While I have you on the phone, let me ask you. You know that nonblister white paint? One of my customers used it on sheet metal. It blistered like hell. What's the story?"

"Let's see, paint. Hmmm, that's Mr. Burgess's department."

"No, no, don't transfer me. Just write it down. Now about this miscellaneous charge on my last invoice. Oh, never mind. Who do I talk to anyway to find out what's going on?"

"Well, sir, I'm sorry, I can really only answer questions about merchandise availability."

"Here's one on merchandise availability. I'm ready to mail 10,000 direct mail pieces. So far I only have sixty of the hundred items listed in the catalog. Should I mail the catalogs or wait for delivery?"

"Uh, let's see. You are a southern customer handled out of our brand-new mechanized warehouse. Our records show all 100 items are in that warehouse."

"What are you telling me? Are the goods lost in the warehouse?"

"Well, sir, it's a new warehouse with some new mechanical and computer systems. It's going to take a little while to get some of the problems ironed out."

"Look, I'm calling long distance. Who can I talk to to get some straight answers fast?"

The expediter was trying to give the customer some straight answers, but he didn't know the answers, so he just made the customer madder. Worse yet, the whole time he talked to the customer he neglected his main job—getting the vendor to ship his company's goods first.

Customer service should answer customer inquiries. Expediting should give customer service periodic reports—statistical and narrative—on delivery status. Customer service can use these reports to answer customer delivery calls. The expediting department should never directly answer customer service inquiries on delivery.

SUMMARY

As long as goods are scarce and people are involved in shipping goods there will be a need for expediters. One professional expediter is worth four clerks recording promised ship dates and truck lines because a professional expediter takes the time to know the vendor's customer service representative personally. The expediter's only real power is his rapport with the vendor's customer service representative. Only through this personal contact can the expediter get the vendor to ship his goods first. Only through this personal approach can the expediter get the vendor to put his order on top of the pile and leave it there.

The expediting department uses a typical expediting system to answer two questions: What are the orders that we need now? Where are the orders now? The expediting system answers the first question by using priority dates on every purchase order the inventory manager issues. The expediting system answers the second question by recording on the float report all orders that have arrived and have not been put away.

Priority dates tell the expediters which orders to expedite first, and the receiving department which orders to put away first. However, the receiving department may have some overriding priorities that force them to handle orders in a certain sequence to maintain their efficiency. The receiving department knows the importance of customer service priority dates, but it is up to the expediters to see that the receiving department moves the goods with priority dates first. Otherwise the receiving department's efficiency priorities will destroy customer service.

Expediting is responsible for insuring the delivery of needed goods and for communicating the status of delivery to customers. Customers should not call expediting directly for delivery information. Expediters are professionals at getting goods from vendors, but they are not professionals at answering customer calls. Expediting should communicate delivery status in the form of narratives and statistical reports to a customer service department.

Insuring the delivery of needed goods completes the fifth step in the focus forecasting inventory management system. The sixth step is measuring performance. When the inventory manager measures performance it gives him the information he needs to change decisions in all the other parts of this system. This feedback process is probably the single most important step in the system. In chapter 9 we will see how the inventory manager measures performance.

Chapter 9

How the Inventory Manager Measures Performance

Measuring performance is the most important part of the finished goods inventory management system. Anything measured improves. Whose performance does the inventory manager measure? Every day he measures the controller's performance by comparing the actual total sales against the controller's total sales budget. Every day he measures his own performance by comparing the actual line-fill rate and months on hand against his inventory plan. Every day he measures the expediters' performance by comparing customer cancellations caused by late vendor delivery and slow put away. He also measures the inventory impact of other departments in the company.

Most of all, the inventory manager measures the buyers' performance. In the end it's the buyers' daily decisions that determine the company's customer service level and inventory turnover. It's the buyers' decisions that determine the company's total gross profit. The controller's sales budget, the inventory manager's plans, the inventory manager's buying systems, the expediting department's procedures for insuring delivery—all these are merely tools that the buyer uses to meet customer service, turnover, and profit goals. Good buyers can overcome any shortcomings in any inventory management system. Bad buyers can make the most perfect inventory management system fail.

THE INVENTORY MANAGER MEASURES PERFORMANCE

Measuring performance is the single most important step in the focus forecasting inventory management system. In 1965 Cleveland Light & Power measured the key indicators of their business by setting up measures to evaluate certain changes in systems and people. Sometimes they changed procedures, sometimes they changed people, sometimes they didn't change anything. But every area they measured improved. They learned that anything measured improves. Buyers' performance in large measure decides the total inventory performance.

There are three measures of a buyer:

1. *Customer service* Are we improving our ability to ship our customers' orders?
2. *Turnover* Are we reducing our inventory investment in relation to sales?
3. *Gross profit* Are we improving our ability to earn gross profit on the things we sell?

If any one of these indicators of performance is measured without the other two, the other two suffer. Measure customer service alone and the smart buyer lowers turnover to avoid the risk of stockouts. Lowering turnover causes more inventory excesses, and the discounting of inventory excesses lowers gross profit. Measure turnover alone and the smart buyer lowers safety stocks to avoid excesses. Lower safety stocks reduce customer service, reduced customer service reduces the company's sales, and reduced sales lower the company's gross profit. Measure gross profit alone and the smart buyer adds new items, increases prices, and limits discounts on inventory excesses. Adding new items lowers customer service and generates inventory excesses; increasing prices reduces sales and inventory turnover; and limiting discounts on inventory excesses reduces turnover and eats up inventory dollars that could be used to improve customer service. The answer is to measure all three—customer service, turnover, and gross profit.

This chapter will show how the inventory manager measures the buyer and others responsible for customer service, turnover, and gross profit. Measuring these three indicators will improve company performance regardless of the inventory management system.

BUYER PERFORMANCE INDICATORS

The inventory manager does not have to set goals for the buyer in the three key areas of customer service, turnover, and gross profit. He only has to compare current performance against other product line segments and against previous periods. If goals are set, the individual buyer should be the one to set them.

In most companies performance measurement means cost control. The accountants keep down the duplicating costs, the telephone expense, and the travel budget. A company's profit-and-loss statement shows that 70 percent to 80 percent of the company's return on investment depends on the cost of goods sold and the size of their inventory. A 1 percent improvement in sales because of improved customer service will pay for all the duplicating costs. A 1 percent reduction in the size of the company's inventory in relation to sales will pay for all the company's telephone expense. A 1 percent improvement in gross profit will pay for the company's entire travel budget. Higher customer service means higher sales. Higher gross profit means higher gross profit percentage on sales. And lower inventory levels in relation to sales mean higher return on investment.

Let's look at a simple report (fig. 9.1) that measures these three areas of a buyer's performance. Figure 9.1 shows that the plumbing buyer has improved in all three areas. Any buyer who improves in all three measures is performing well. Any company that improves in all three measures is a healthy company. Customer service, inventory turnover, and gross profit are the key measures of a buyer. They are like a three-legged stool. When one leg is shorter than the other two, the stool falls over. When all three are balanced the company can't help but prosper. Service attracts sales, sales help turnover, and turnover with reasonable gross profit means higher return on investment. Don't worry that the buyer travels too much, that he can't communicate, that his expense account is out of line. Just see whether he is improving customer service, turnover, and gross profit and forgive anything else.

Figure 9.1 BUYER PERFORMANCE INDICATORS FOR JULY, THIS YEAR

Product line	Customer service		Inventory turnover		Gross profit ($ millions to date)	
	July, last year	July, this year	July, last year	July, this year	Last year	This year
Giftware	87	88	3.7	3.5	409	449
Sporting goods	84	84	3.2	3.4	342	330
Paint	91	85	3.8	4.0	197	218
Building supplies	82	85	4.0	4.0	87	117
Fasteners	89	87	4.0	3.8	342	313
Lawn, garden	78	74	4.2	4.5	485	304
Tools	85	87	4.1	4.0	386	414
Electrical	91	92	3.7	3.5	207	286
Plumbing	94	94	3.8	3.9	259	307
Total	87	88	3.9	3.7	2,714	2,738

Let's look at each one of these key buyer performance indicators in more detail. First let's look at customer service.

CUSTOMER SERVICE

In this finished goods inventory management system the measure of customer service is line-fill rate. Line-fill rate is the percentage of lines a customer receives of the total lines ordered. If on the average we ship 90 of every 100 lines ordered our line-fill rate is 90 percent. What about the other 10 percent? Why did we fail to ship those lines? There are four major reasons for failing to ship a customer's order for an item:

1. Our goods are available but we failed to ship them.
2. Our goods are at our receiving dock but not put away for order filling.
3. The vendor or our plant has failed to deliver on time.
4. We didn't order enough.

The inventory manager and the buyer are responsible for the line-fill rate. Other departments are responsible for some of the reasons for failing to ship a customer's order.

Operations Department Quality

Our goods are available but we failed to ship them. Why? Operations departments are like factories; we usually measure their productivity: (a) dollars shipped per employee, (b) pounds shipped per employee, (c) lines shipped per employee, (d) total cost as a percentage of sales. We encourage speed, not quality.

Measure dollars shipped per employee and he won't travel far to pick up two nuts and bolts. Measure pounds shipped per employee and he may skip a plastic swim pool order. Measure lines shipped per employee and he may not fetch a ladder to pick a gross of blue ball gloves from an upper shelf. Measure the total operations cost as a percentage of sales and the operations manager will shut his eyes when employee fails to ship a customer's order.

The first key to better customer service is to measure quality as well as speed. The inventory manager knows the on-hand inventory balance for each item. How many customer order lines did we fail to ship when the inventory manager's computer records indicated we had stock on hand? This measure tells how many customer order lines we canceled when our goods were available but we failed to ship them.

Operations Department Speed

Our goods are at our receiving dock but not put away for order filling. The second key to better customer service is to measure the operations department's speed in moving needed goods from the receiving dock to the shelf. Few things are more frustrating than knowing the goods are here but can't be shipped because they haven't been put away. An operations department has a certain number of people and a certain amount of work, so they set priorities of their own, usually (a) pick customers' orders, (b) load outbound trucks, (c) unload inbound trucks, (d) put away goods received, (e) sweep the floor. It's clear that customer orders must be picked. Orders picked are sales and sales against operations expense is a basic measure of warehouse efficiency. It's just as clear that outbound trucks must be loaded. Until the goods are shipped, they can't be billed. It's also clear that inbound trucks must be unloaded.

Inbound trucks sitting at the warehouse dock add demurrage to warehouse expense.

But what about put away? It's so intangible. The only time it takes higher priority is when the receiving department is so jammed that they can't unload inbound trucks. Seldom does an operations manager use overtime to put goods away. Overtime is a real expense and shows on the financial statement. Put away, that's another matter. Canceled lines are the buyer's problem. The expediting system can tell us how many customer order lines were canceled because the goods were not put away.

Expediting Department

The vendor or our plant has failed to deliver on time. The third key to better line-fill rate is reliable delivery. The expediting system considers customer order cancellations a delivery problem if the vendor's order arrives after the promised delivery date. Many companies pride themselves on hard-nosed price negotiations. Some better companies measure quality as well as price. Few companies select a supplier based on delivery performance. Yet the primary reason for changing to a new supplier is failure to meet delivery schedules.

Generally less than 4 percent of the suppliers have 50 percent of the customers' line cancellations. How many customer order lines were canceled because of slow delivery? The expediting system knows the expected delivery date of every order. The expediting system can keep track of how many customer order lines were canceled because the supplier failed to deliver on time. The expediting system measures the expediting department's ability to insure the delivery of needed goods.

Buyers

We didn't order enough. The last key to improving line-fill rate is to count the number of customer order lines that were canceled because we didn't order enough. The reasons we didn't order enough fall into three categories:

1. *Adding too many new items.* A company must have new items to

grow, but new items are risky. New items must be added at a pace consistent with turnover and fill-rate goals. Buyers are greedy. If a white wall clock sells well, put in a green one. If both wall clocks sell well, put in an avocado model. New items hurt customer service and decrease turnover. There's no demand history. How can the buyer buy right? The number of lines canceled due to new items increases with the number and speed of new items added to the company's line.

2. *Adding too many allocated items.* Some goods are scarce because of production availability. Buyers do not decide how many of these items to buy, they only decide how much of an allocation to take. The buyer must consider the delivery impact of adding allocated items. Having too many allocated items in the company's line prevents the inventory manager and the buyer from improving customer service.

3. *Not buying enough to support customer orders.* The inventory manager is responsible for developing strategies that make a reasonable forecast of future demand. Sometimes the buyer does not buy enough because the focus forecasting sales forecast is low. Sometimes the buyer does not buy enough because he believes the focus forecasting sales forecast is high. The buyer can change any suggested buy, but he is responsible for cancellations when he didn't order enough.

The inventory manager knows when an item is not on hand, when an item is not at the receiving dock, when the supplier has not exceeded the average or promised lead time. Thus the inventory manager can determine how many customer order lines were canceled because the buyer didn't order enough. The finished goods inventory management system shows the percentage of customer order cancellations by reason on a customer cancellation analysis report.

Figure 9.2 is a customer cancellation analysis showing the number of lines canceled for each item. Next to the item is the reason for the cancellation. At the bottom are the total lines canceled by reason. In this example, in 19 percent of the cases our goods were available but we failed to ship them. In 18 percent of the

Figure 9.2 CUSTOMER CANCELLATION ANALYSIS

Item number	Item description	Number of lines	7/22/this year Reason for cancellation
		Other items	
7438	2 × ¾" valve	60	IF
7439	Frost cinaware	100	AL
7440	30' extension ladder	3	OH
7441	16 HP mower	2	LT
7442	TV video game	7	NI
7443	Weather strip tape	18	UB
7444	Weather strip tape	27	UB
7445	10P nails	14	IF
		Other items	

Total Cancellations by Reason

Number of cancelled lines	OH (On hand)	IF (In float)	LT (Lead time)	NI (New item)	AL (Underbought allocated)	UB (Other)
2000	380	360	420	300	60	480
% distribution	19	18	21	15	3	24

cases our goods were at the receiving dock but were not put away. In 21 percent of the cases the vendor or our plant failed to deliver goods on time. In 42 percent of the cases we didn't order enough. Of the 42 percent where we didn't order enough, 15 percent of the cancellations were on new items, 3 percent were on allocated items, and we simply didn't order enough of the other 24 percent.

The inventory manager measures the total cancellations by reason every day. Cancellations by reason direct the inventory manager to the department that's causing the failure to meet line-fill rates. Remember that anything measured improves. Each department responsible for the line-fill rate pays careful attention to the distribution of cancellations by reason. This measure is a key to improving overall customer service.

We've seen how the inventory manager measures the buyer on customer service. Now let's see how the inventory manager measures the buyer on turnover.

TURNOVER

The inventory manager measures turnover in terms of months on hand. Months on hand is the number of months the inventory can support future sales without replenishment. Classic inventory turnover is the annual sales divided by the average inventory investment. In either case, turnover is the inventory investment in relation to sales.

There are only two ways to improve turnover, increase sales or reduce inventory.

Increasing Sales

By the 80/20 rule, 20 percent of the items in an inventory generate 80 percent of the sales. There's an interesting corollary that says 20 percent of the customers account for 80 percent of the company's sales. So one of the simplest ways to increase sales is to sell the top 20 percent of the items to the top 20 percent of the customers. This principle is called distribution analysis. The distribution analysis shows each one of the major customers and the best-selling items they are not purchasing.

The distribution analysis in figure 9.3 shows that Mark Lumber is not buying our second-best-selling item in the whole company. Mark Lumber would be delighted to know which best-selling items it is not purchasing.

Distributing the best-selling items is the simplest way to increase sales. The inventory manager and the buyer are not responsible for the distribution of the best-selling items in the company, they are responsible for inventory turnover. The inventory manager should make sure that the sales department is distributing the best-selling

Figure 9.3 DISTRIBUTION ANALYSIS

Mark Lumber		
Item	Description	Rank
5436	20 gal. corrugated garbage can	2
5706	48 pk 60 watt bulb	7
1297	White semigloss gals	14
3046	5 gal. gas can	17

items in the company to the company's best customers. This is the simplest way to improve sales without adding items to inventory.

There are some steps the inventory manager and the buyer can take to increase sales. The buyer should add and discontinue items based on sales potential. Of course, with many new items this buying decision is more art than science. Still, the company's sales records can guide the buyer's selection. Ideally the inventory manager can classify item sales by end use. The buyer can then look at the average sales per item within a classification. The higher the average sales per item for a classification, the more likely additional items will generate significant sales in that classification.

Looking at figure 9.4, the buyer can see that only four tape players are generating $36,000 a year in sales. The buying question is then how to add tape players without splintering the sales volume of the existing four items. Does the buyer have good, better, and best tape players? Does the buyer have an eight-track tape player and a six-track tape player? Does the buyer have a car model tape player? Would customers who buy Panasonic tape players also buy American brand tape players? There are probably tape players the

Figure 9.4 GIFTWARE DEPARTMENT

Fine line classification	Annual sales	Number of items	Average annual sales per item
Irons	$ 90,100	10	9010
Tape players	36,000	4	9000
Dinnerware	84,150	12	7013
Radios	40,300	12	3358
Openers	52,000	18	2888
Vacuums	30,900	12	2575
Clocks	84,050	35	2401
Televisions	40,850	18	2269
Mixers	16,700	9	1855
Ice crushers	10,750	8	1344
Blenders	11,650	10	1165
Popcorn poppers	2,200	4	540
Ice cream makers	1,150	3	386
Total	$500,800	155	3236

buyer can add without hurting sales on existing tape players.

What about the thirty-five clocks? Are they all needed to provide selection? Or should some of the slower-selling clocks be eliminated to hold down the total number of items in the giftware department? Eliminating slow-moving items frees inventory investment to add fast-moving items. Adding best sellers improves sales in relation to inventory investment.

Besides adding items based on sales potential the buyer can do two other things to improve sales. The buyer can improve customer service. Generally speaking the buyer can improve customer service only by increasing inventory investment, but increasing inventory investment lowers inventory turnover. The buyer can also improve sales by reducing gross profit, since reducing gross profit on an item makes the item more attractive to the customers. But reducing gross profit to improve sales lowers the department's total gross profits and perhaps the buyer's overall gross profit performance. There are all sorts of marketing and sales analysis tools to help the buyer improve sales, but improving sales is still more art than science and it's up to the individual buyer to improve sales without lowering inventory turnover or reducing gross profit performance.

Reducing Inventory

The second way to improve turnover is to reduce inventory. The prime target for reducing inventory is reducing excess inventory. Everybody gets some excess inventory sometimes. The first step, of course, is to identify the excesses. It's important to do so objectively and to pick out the most significant inventory excesses. There may be a two-year supply of corks on hand and only a one-month supply of garbage cans. But a two-year supply of corks is probably only $14 too many corks, while a one-month supply of corrugated garbage cans may be $10,000 too many.

The key to selecting the most significant excesses is to pick excesses that are significant in dollars and significant in months on hand. Figure 9.5 shows a monthly excess inventory listing for sporting goods, by vendor. For each of these items we have more than six months on hand and over $500 of excess. The buyer can take four steps to reduce inventory excesses:

Figure 9.5 INVENTORY EXCESS LISTING: WAREHOUSE C

Vendor number	Vendor name	Item number	Item description	Excess investment ($)	Quantity on hand	Excess quantity	On order	Action
							July 1, This Year Over 6 months on hand Over $500 Excess	
6691	Olympia	10389	10-speed bike	25,850	670	470		
		10491	5-speed bike	8,400	180	240		
6893	Tender Care	25382	14" live trap	6,900	2930	2300		6/1 20% off
		25383	16" live trap	1,300	3050	350		6/1 20% off
		25384	18" live trap	4,200	2400	824		6/1 20% off
6901	Kusuki	20687	Motorcross bike	600	10	2	3	
7023	Mountain Seed	27468	Wildflower seed	880	5584	1000		
..	..	..	..	..	..	..	..	..
..	..	..	..	..	..	..	..	..
Total				342,800				

1. Cancel open orders.
2. Transfer excesses between warehouses.
3. Return excesses to the supplier.
4. Discount excesses.

The buyer writes the step he will take on the inventory excess listing, and this step is recorded on the next inventory excess listing. The next inventory excess listing shows how well the buyer's action reduces excess inventory.

The months-on-hand and dollar amounts shown in the inventory excess listing are based on focus forecasting's projection of the items' future sales. But an item can quickly move from too little inventory to an inventory excess at the end of the season. Let's look at the four actions that a buyer can take to reduce excess inventory in more detail:

1. *Cancel open orders.* At the end of a season the company really scrambles to get in needed goods. Immediately after the season is over these same needed goods quickly become inventory excesses. The easiest way to avoid inventory excesses is to quickly cancel open purchase orders at the end of a season.
2. *Transfer excesses between warehouses.* A multiple-warehouse company can redistribute excess inventory. The best way to get rid of excess inventory once it occurs is to transfer it where it is needed. At the end of every month the inventory manager runs an inventory transfer report to redistribute inventory excesses. The buyers review the inventory transfer report to decide which items they will actually transfer.

Figure 9.6 shows the logic for redistributing inventory excesses. Warehouse C has 6.7 months on hand. The total months on hand for the whole company is 4.5 months on hand. The inventory transfer calculation redistributes the total inventory so that each warehouse has 4.5 months on hand. When the buyer transfers the excess inventory from warehouse C to warehouse A he eliminates the possibility of buying 10-speed bikes for warehouse A while he has an excess inventory of 10-speed bikes in warehouse C.

Figure 9.6 REDISTRIBUTING INVENTORY EXCESS

Item—10–speed bike	Warehouse			
	A	B	C	Total
Months on hand	2.4	4.3	6.7	4.5
On-hand quantity	240	215	670	1125
Forecast sales Distribution per month	100	50	100	250
4.5 months on hand	450	225	450	
Current on hand	240	215	670	
Transfer quantity	+210	+ 10	−220	

Inventory transfers are costly. The inventory manager limits the number of inventory transfers through parameters given in the inventory transfer report. These parameters limit transfers to items with more than a certain number of months on hand, more than a certain number of dollars of inventory excess, more than a certain number of units being transferred, more than a certain number of dollars being transferred, and to items needed in one of the other warehouses. The inventory manager varies these parameters to generate full truckloads between warehouses to make sure the transfers between warehouses are economical. These transfers are the most profitable means of disposing of inventory excesses.

3. *Return excesses to the supplier.* Inventory excesses are frequently caused by suppliers' errors. Sometimes suppliers overship purchase orders, sometimes they ship the wrong item. Even when the inventory excess is not caused by the supplier, the supplier may be willing to accept a return, especially if he has a short supply of the item. Usually a supplier will accept a return in exchange for a handling charge. If an inventory excess cannot be used within the company, returning it to the supplier is usually the most profitable means of disposing of it.

4. *Discount excesses.* The fourth way to dispose of inventory excesses is by discounting. The buyer's first discount should be the only discount. It should be big enough to move the goods into and out of the customers' stores. Discount bulletins should

be issued on a regular schedule and they should offer a group of items so that customers can review the offering and make one low-cost decision. The discount bulletins should be limited to the quantities that the buyer considers excess.

There are customers who make a living disposing of distress merchandise. They will usually buy an assortment of distress merchandise regardless of the content for a price. The smart buyer has a regular distress merchandise customer who will buy excess inventory, one big enough to handle the volume and solid enough not to be a credit problem. Ideally, the distress merchandise customer should not compete with the buyer's regular customers. The section on improving gross profit gives some rules for determining discounts.

To summarize: To increase turnover there are only two things the buyer can do.

1. *Improve sales.* Distribute the company's best-selling items especially to the company's best customers. Add items in the best-selling categories. Discontinue poor-selling items. The buyer decides which items to add and discontinue. This decision is often more art than science.
2. *Reduce inventory.* Identify excess inventories. Quickly cancel open purchase orders at the end of a season. Transfer excess inventories between branches. Return excesses to the supplier. Discount firmly. Sell groups of items and limit sales to excess quantities.

The inventory manager measures the buyer's turnover, the buyer's ability to improve sales and reduce inventory. The buyer's decisions to improve sales and reduce inventory affect customer service and gross profit. We have already seen how the inventory manager measures the buyer's customer service. We've seen the steps the inventory manager takes to help the buyer improve customer service. We have already seen how the inventory manager measures the buyer's turnover. We have seen the steps the inventory manager takes to help the buyer improve sales and reduce inventory. Now let's look at how the inventory manager measures the buyer's gross profit performance.

GROSS PROFIT

Gross profit is the difference between the selling price and the cost of goods sold. Gross profit is the money that the company earns for taking the risk of investing in the inventory necessary to sell a product. The higher the risk, the more gross profit the company should earn. This is true of all investments. If you invest your money in a bank account you receive a relatively low interest rate because you are taking a very little risk. If you invest your money in the stock market you expect a higher return on your investment because you are taking a greater risk.

When a company invests in product inventory, their risk depends on the item. Some stable products with high turnover involve very little risk. On these items a company can afford a relatively low gross profit since the return on its inventory investment is a very low risk. Seasonal items require a higher degree of risk because there are more leftovers and lower inventory turnover. Companies should earn a higher return on their seasonal inventory investment because of this higher risk. Items that must be bought in full truckloads result in low turnover and are thus higher risks. When a company must buy full truckloads of merchandise to keep an item in inventory, it should return a higher gross profit on the inventory investment.

When you invest in anything, you look at the return on your investment . . . 5%, 6%, 8%, etc. When the company invests in inventory, the company should look at the return on inventory.

RETURN ON INVENTORY

Return on inventory is the gross profit rate times the inventory turnover. Take the example in figure 9.7. *Step 1:* The gross profit rate is the annual gross profit divided by the annual sales. *Step 2:* The inventory turnover is the annual sales divided by the average inventory investment. *Step 3:* The return on inventory is the annual gross profit divided by the average inventory investment. *Step 4:*

The return on inventory is also the gross profit rate times the inventory turnover.

Figure 9.7 RETURN ON INVENTORY CALCULATION

Step 1: The gross profit rate	
Annual sales	$1,000,000
Annual gross profit	160,000
Gross profit rate	16%
Step 2: The inventory turnover	
Annual sales	$1,000,000
Average inventory investment	250,000
Inventory turnover	4
Step 3: The return on inventory	
Annual gross profit	$160,000
Average inventory investment	250,000
Return on inventory	64%
Step 4: The return on inventory	
Gross profit rate	16%
Inventory turnover	4
Return on inventory	64%

The buyer must improve total gross profit. The inventory manager measures the buyer's performance on customer service, inventory turnover, and gross profit and combines these three measures in the return on inventory analysis.

Figure 9.8 shows a return on inventory analysis for the sporting goods buyer. This analysis provides early indications of possible future gross profit problems. It shows the gross profit content of open purchase orders, on-hand inventory, and demand to date. It shows the gross profit lost because of discounts to move excess inventory. This analysis shows the turnover lost because of customer order cancellations. It shows the total inventory turnover, the return on inventory, and the total gross profit dollars.

The company has an overall return on inventory goal of 64 percent. The sporting goods inventory turnover goal of 4 is based on past experience. Their gross profit goal of 16 percent is the company's return on inventory goal of 64 percent divided by the sporting goods inventory turnover goal of 4. To meet the total gross profit dollar goal, the sporting goods buyer must either raise the gross profit rate to 16 percent or make up the lost gross profit dollars in higher customer service or inventory turnover.

Let's look at figure 9.8 column by column.

Column 1 is the initial gross profit rate for goods on order. It is the earliest indicator of future gross profit.

Column 2 is the initial gross profit rate for goods on hand. This rate indicates the near-future gross profit.

Column 3 is the initial gross profit rate for demand to date. If it were not for discount sales, this would be the final gross profit rate.

Column 4 is the gross profit rate lost through discounting excess inventory.

Column 5 is the final gross profit rate on demand to date.

Column 6 is the initial inventory turnover. It's the demand to date annualized and divided by the average inventory investment.

Column 7 is the initial turnover lost through customer order cancellations to date.

Column 8 is the final inventory turnover. It's the sales to date annualized and divided by the average inventory investment.

Column 9 is the return on inventory rate. It is the final gross profit rate times the final inventory turnover.

Column 10 is the gross profit dollars to date. It is the final measure of the buyer's ability to maintain gross profit rates, reduce customer order cancellations, and improve inventory turnover.

The inventory manager measures the buyer's performance on customer service, inventory turnover, and gross profit. The return on inventory analysis measures the buyer's gross profit performance. The return on inventory measure makes sure the buyer maintains customer service and inventory turnover as he works for higher gross profit dollars.

SUMMARY

The inventory manager measures customer service, inventory turnover, and gross profit. He measures the buyer's performance, and in the process he measures his own performance and the performance of the other managers in the company.

Figure 9.8 RETURN ON INVENTORY ANALYSIS

	1	2	3	4	5	6	7	8	9	10
						Sporting Goods				
	Initial Gross Profit (%)			Discount sales loss to gross profit (%)	Final gross profit (%)	Initial inventory turnover	Turnover lost through cancellations	Final inventory turnover	Return on inventory (%)	Gross Profit to date ($000's)
	goods on order (%)	goods on hand (%)	demand to date							
Jan	14.7	15.8	15.7	−0.7	15.0	3.5	0.4	3.1	46.5	804
Feb	14.6	15.6	15.4	−0.3	15.1	3.8	0.7	3.1	46.8	1560
Mar	15.0	15.3	15.2	0.0	15.2	4.1	0.6	3.5	53.2	2260
Apr	15.5	15.4	15.3	−0.1	15.2	4.2	0.6	3.6	54.7	3160
May	16.0	15.7	15.5	−0.2	15.3	3.9	0.3	3.6	55.1	4160
June	16.2	15.9	15.7	−0.2	15.5	3.8	0.2	3.6	55.8	4960
Goal					16.0			4.0	64.0	$5780

To improve customer service the inventory manager determines the reasons for canceled customer orders. Sometimes available goods were not shipped. Sometimes the goods arrived but were not put away. Sometimes the vendor ships late, and sometimes the buyer just didn't buy enough. The inventory manager knows the reasons for cancellations and measures the operations department, the expediters and the buyers to improve performance.

There are only two ways to improve inventory turnover: improve sales or reduce inventory. One way to improve sales is to distribute the company's best-selling items to its best customers. Another way is to replace slow-moving items with fast-moving items. The best way to reduce inventory is to get rid of excesses. Cancel end-of-season orders quickly. Transfer excesses to warehouses that need them. Return excesses to the supplier. Discount excesses firmly and early.

Companies earn gross profit for taking a risk. Since lower turnover items are riskier, they should provide a higher gross profit. The return on inventory analysis can measure gross profit performance and its impact on customer service and inventory turnover. The return on inventory analysis can give early indications of future gross profit problems. The gross profit on order, on hand, and in demand give these early indications.

Customer service, turnover, and gross profit are the three key measures of a buyer and of the inventory manager. When a company improves all three, it can't help but prosper. Customer service attracts sales. Sales increase turnover. And turnover with reasonable gross profit is return on investment.

Now we have seen the six steps in this finished goods inventory management system:

1. The controller budgets total sales.
2. The inventory manager plans total inventory levels.
3. The inventory manager sets total safety stock levels.
4. The buyers buy individual items.
5. The expediters insure delivery of needed goods.
6. The inventory manager measures performance.

My company uses this system. It's a simple system; people understand it, and they make it work. The results have been outstanding. Our forecasted demand varies from actual demand by less than 1 percent. Our buyers change less than 8 percent of our computer's buying decisions. At the end of this fiscal year our inventory was within 0.5 percent of our planned level. Our service level was the highest in the history of our company.

This is the finished goods inventory management system that focus forecasting is part of. This is how to keep stock.

Chapter 10

Seven Steps in Creating New Systems

I have talked to about 2000 people around the country about focus forecasting. They usually get pretty excited about it. I get telephone calls saying, "Bernie, we're about to put in a new forecasting system. Can you tell us the seven strategies in focus forecasting?" I'm reluctant to tell them what the seven strategies are because I think they are asking the wrong question.

Inventory management is a balancing act. A company's existing system may not be perfect, but it's kept the company in existence. Tampering with that delicate balance without planning can be disastrous. I've been creating new systems since I was sixteen. Just the other day I went with my wife and children to Washington D.C. I felt very old when I saw the computer I used to program exhibited in the Smithsonian Institute. One thing all these years in systems have taught me is before you do anything, know why you are going to do it.

I remember visiting a company in Bristol, Connecticut, that had just installed a new material requirements planning system. At least they thought they had. The company had 30,000 raw materials. The system spit out so much information on each item that it took three lines to print it on the 14″ wide computer form. To make it "easy" to read, the first line on the form was chicken-egg brown, the second line was pistachio green, and the third line was white. The computer printed very fast. The computer operator looked hypnotized: brown, green, white; brown, green, white; brown, green, white. Now and then the operator would wheel in another box of three-part forms and wheel away the mountains of paper that were already printed.

"How do your purchasing people like this report?" I asked the data processing manager.

"Well, they're not using it yet. They want it by purchasing agent and we don't have the materials coded yet. They want us to code the items. They say they don't have time. Boy, what we need around here is some top management support. We need somebody to make those purchasing agents get involved."

"You mean you print this report and nobody uses it?" I asked.

"No, we use it. We check it for errors. We make sure the files are being updated right. We're going to make it a six-part form and print it twice a week. That way every purchasing agent can get a copy at least once a week."

I stood there watching that stream of brown, green, and white pour through the computer. I knew the last thing this data processing manager needed was a theoretical discussion. He was probably going to get fired. I knew that if our purchasing system worked like this one I'd get fired.

"Well," he said, "what would you like to know? This is a multilevel bill of material explosion. We can tell you every material an item uses. But it doesn't end there. We have upward chains as well as downward. Pick a material and we can tell you every item that uses it."

I thought for a minute and said, "Why do you run this system at all? What value does it have if the purchasing agents don't read the reports?"

His answer scared me because it tempted me to use this program for the same reason. He said, "Just because those purchasing agents aren't up to using this system doesn't mean it isn't valuable. Do you know I'm one of only three people in the state who can set up this program? I get two or three job offers every week. It may not be of value to those purchasing agents, but it sure as hell is to me."

Focus forecasting is a good tool. The finished goods inventory management system is a good tool. But nobody should use tools without knowing why. Here are the steps you can take to create new systems. The first few steps help you know why you want a new system.

STEP 1: MEASURE THE PERFORMANCE OF THE EXISTING SYSTEM

There is always an existing system; people do things routinely by nature. The best place to start creating new systems is to measure the performance of the existing system.

Customer Service and Turnover

Go back as many years as possible and develop a history of customer service and turnover year by year, month by month. You'd be surprised what the actual figures are as opposed to what people think they are. If there aren't any records, start measuring what the customer service and turnover is now. Without some starting point it's impossible to tell whether new systems are improving things or making them worse.

Ordering Costs and Inventory Carrying Costs

Measure the lines of work that the existing systems generate now. If a company has 25,000 items and they order once a month they have 300,000 lines of work. Relate the lines of work to the inventory size. If the company does 300,000 lines of work and has $3 million in inventory, a dollar of inventory is worth 0.1 lines of work. If the new system improves customer service while reducing the size of the inventory and reducing the lines of work, you know it's working right.

Forecasting Accuracy

If the company doesn't have a formalized forecasting system you can't measure forecast accuracy. You can only measure results. But many companies do have existing forecasting systems. Measure the overall forecast accuracy. How did the total dollars forecast compare with the actual dollars of demand? Classify the items into volume segments. What's the forecast accuracy on the best-selling items? How is it on the worst-selling items? Is the forecast consistently high or low? What's the range of the forecast error? Is it 100 percent wrong sometimes? These measures let you know whether any new forecasting system is really improving forecast accuracy.

Excess Inventory

Add up the dollars you have invested in inventory when the on-hand inventory is more than the reorder point. Find out how much profit is lost by discounting inventory to eliminate excesses. The new system should reduce the number of inventory excesses and the total dollars of excess. The new system should reduce the profit loss through discounting to get rid of excesses.

This is the beginning point for creating any new system: How are you doing now? If you don't know how you are doing now, you can't know whether you are doing any better with a new system. There's an advantage in just setting up these measurements. Anything measured improves. Just the process of setting up the measurements improves performance.

STEP 2: SET REALISTIC GOALS

Many companies have customer service and inventory turnover goals. These are the stated goals that they tell their customers and their employees. Find out whether the goals are realistic. Has your company in its best years ever met its stated goals? If not, the goals are certainly unrealistic now. You can use some logic to see what the goals should be.

You can have the computer simulate past experience. The computer can tell you what level of safety stock you need to reach the company's stated customer service objectives. Suppose the computer tells you that on the average you need 1.5 months of safety stock to meet the company's stated customer service objectives. You can then see how often the company reviews items for buying decisions. If the company reviews items for buying decisions once a month on the average you should have 0.5 month's supply for review time inventory investment. Ideally, then, your company should have 1.5 months of safety stock and 0.5 months of review time. Ideally you would have 2 months of supply.

You can adjust the ideal inventory goal to arrive at a reasonable inventory goal. Look at the items. Are there a lot of new items? Are there a lot of seasonal items? Is the company taking advantage of special buying opportunities? Are the vendors inconsistent in their delivery? You can judge what the reasonable excess inventory would be. Suppose you decide the excess inventory should be 0.5 month's supply. You should add the excess inventory to the safety stock inventory investment and the review time inventory investment to arrive at a reasonable inventory goal. In this example you would add 0.5 month's excess inventory to 1.5 month's safety stock and 0.5 month's review time to arrive at a reasonable inventory goal of 2.5 month's supply. Of course, this inventory goal would be reasonable only if the company's stated customer service objective was reasonable.

You must review these inventory goals with the company operating management and with the company's general manager. Reasonable inventory goals become realistic inventory goals only if your company management accepts them. You must fight to make the company's stated goals realistic right from the beginning. You must have realistic inventory goals before any new systems are started, for without them, you can never win.

STEP 3: FIND THE CAUSES OF THE PROBLEM

Compare the realistic inventory goals against the actual performance. The four major reasons for not meeting customer service goals are:

1. Our goods are available but we failed to ship them.
2. Our goods are at our receiving dock but not put away for order filling.
3. The vendor or our plant has failed to deliver on time.
4. We didn't order enough.

We have to find out which of these reasons are causing the major part of our customer service problems. For example, if the major problem is failure to ship available goods, we probably need some new warehouse systems. We probably need a system to improve our order-filling accuracy. We probably need more measurement and discipline on our order fillers.

If the major problem is that goods at the receiving dock have not been put away, we probably need some better receiving systems. Do we have a method of telling the warehouse people which orders are needed first? Do we have a way of measuring their put-away performance on a daily basis? Are we budgeting the people properly? Can we accurately forecast what the inbound shipments from vendors will be?

If the major customer service problem is that the vendor or our plant has failed to deliver on time, we may need some work in expediting. Do we have an expediting staff? Have we set up distribution requirements planning systems with our major suppliers? Are we estimating the average vendor lead time properly in the first place? Are we using material requirements planning systems to control the movement of goods through our plants?

If the major part of our customer service problem is that we didn't order enough, we must ask whether our buying systems are effective. Are we being realistic about the amount of inventory we need to meet our stated customer service goals? Are there too many new items in the inventory mix? Are our forecasts of item demand accurate? Are our safety stock strategies working properly?

Finding the reasons for failure to meet customer service goals provides the direction for new systems development.

We can also determine the reasons for excess inventory on different items in our company. Are we building excess inventories because we can't resist making such a deal with a supplier? because our forecasts of the future are inaccurate? because our freight minimums are too high? because we have added too many items to the company's inventory? Answering each of these questions can provide direction for new systems development.

Of course, this same analysis may point out that we don't need new systems. It may reveal a need to change company philosophy. It may show that certain people are not doing their job. Whatever it shows, it's extremely important to find the causes of the problem. Once we know the causes, we can be certain that whatever system we undertake is headed in the right direction. We can attack the most significant problem and work in the area that needs the most improvement. It's just common sense to find the causes of the problem before undertaking new systems development.

We've measured the performance of our existing system. We've set realistic goals. We've found the causes of our problems. Now we know we need a new system and we know why. The next step is just as important as the first three. We have to involve the people who will use the system.

STEP 4: INVOLVE THE PEOPLE

If people are part of creating a system, they will use it. Let the people who are going to read the reports design them. In every system someone is doing something at some time. These who, what, and when's are procedures. Wherever possible let the people who will follow the procedure write the procedure. Let the people who will use the system design the system. That's what involve the people means.

When I first worked on focus forecasting, I asked the buyers "How do you forecast?" They gave me the simple strategies. They were happy to give them to me. They were part of the very beginning of creating the new system they would use. When we first tested focus forecasting with computer simulation we played "Can you outguess focus forecasting" with the buyers, and they beat the computer forecast. But each time, they made focus forecasting better; they became more involved in creating the system. Focus forecasting is their system; they created it, and they make sure it works.

Some systems designers have a real pride of authorship. They don't want anybody to see the system until it's perfect. These people are creative and intelligent, but they are missing the secret to creating systems that work. A professional knows the "art of noninvention." A professional knows that the only way to create systems that work is to involve the people who will use the system.

STEP 5: TEST THE NEW SYSTEM

Try out the new system on a test basis. Forecast past demand. Use the buying strategies on items that you know are problems.

Calculate priority dates on purchase orders and see whether they make sense. Run some reports on past data, then let the people look at them. If the system can't work on the past data, it doesn't have a chance of working in the future. Let the users evaluate the test results. It's another chance to involve them.

Before focus forecasting ever made a real buying decision, we had tested its effectiveness on thousands of items. Appendix 13 shows twenty items with their two-year demand patterns by month taken directly from a focus forecasting simulation run. At the bottom of the simulation run is a summary of simulated forecast errors, months on hand, and customer service for the past six months. Every time we change focus forecasting, we measure the impact on the past six months' history on over two thousand items because we know that if we can't improve performance on what has already happened, we have little chance of improving future performance.

STEP 6: TURN IT ON

On a certain date you will replace the old system with the new system. New systems always have bugs, problems no one expected. As long as the people who use the system were involved from the beginning, bugs are minor problems. People involved from the beginning will make the new system start up smoothly. Focus forecasting still has bugs, but they don't stop the buyers from using the system effectively. They see a bug as an opportunity to improve our overall performance.

I often hear systems people saying, "The system would work if our people had some discipline." Or "The system would work if we had top management support." Sometimes they say, "Sure our system has some bugs, but our people make a mountain out of a molehill." They need to remember to involve the users. If it's their system, they will make it work.

STEP 7: MEASURE THE PERFORMANCE OF THE NEW SYSTEM

The last step in installing a new system is to measure its performance. Is customer service getting any better? Have we reduced the size of inventory in relation to sales? Have we reduced the overall lines of work in the company? We can go through the seven steps of creating a new system as many times as we have to. As long as we are measuring performance each time we add a new system or change a system, we'll know whether we are making things better or worse. So you can see that measuring performance is the first and last step in the creation of new systems.

SUMMARY

There are many powerful, sophisticated inventory management tools available. Before you use any of them you should know why you are using them.

This chapter talked about seven practical steps for creating new systems:

1. Measure the performance of the existing system.
2. Set realistic objectives.
3. Find the causes of the problem.
4. Involve the users.
5. Test the system.
6. Turn the system on.
7. Measure the performance of the new system.

The two most important steps in this procedure are measuring performance and involving people.

The inventory manager and the data processing manager are the ones primarily responsible for creating the finished goods inventory management systems. The next two chapters tell you the role of the inventory manager and what to expect from data processing.

Chapter 11

The Role of the Inventory Manager

Whenever I look at a business problem, I pretend I'm the sole owner of the company. I pretend it's a little company so that I can get at the fundamentals of the problem.

The owner of the company has an inventory problem. He wants better customer service. It's annoying and costly when customers call about out-of-stock conditions. He wants lower inventory investment because inventory investment chews up the cash that he needs to continue to grow. He wants lower ordering costs. He knows he is losing freight and volume discounts. He knows it is costly to handle small orders. The owner knows there are systems to improve service, lower inventory, and reduce ordering costs.

If I were that company's owner, here is how I would go about creating an inventory management system for my company.

1. *Find the right inventory manager.* The right manager needs a winning record to survive his first wrong decision. The inventory manager must be an expert in inventory management systems. He won't have time to reinvent the wheel. The inventory manager must be human. He must see his job in terms of getting things done through people. The right inventory manager may already be in the company.
2. *Set realistic goals.* In most companies there is a big difference between customer service policy and customer service reality. The inventory manager must set realistic short-term goals and must work with the company president to set realistic long-term goals.

3. *Plan specific objectives.* Without specific objectives, goals are hard to achieve. The inventory manager must use his own knowledge and the knowledge of others in the company to plan inventory objectives that lead to the company's long-term goals. These objectives must be specific and measurable.

With the right inventory manager, realistic goals, and specific objectives, the company can start to solve its inventory problems. People, goals, and plans create inventory management systems that work.

First let's see how to find the right inventory manager.

FIND THE RIGHT INVENTORY MANAGER

There are all kinds of inventory managers. Some are systems people, some are production people, some are operations research people, some are financial people, some are even college professors. Successful inventory managers have one thing in common—the people in their company believe in them. They have confidence the inventory manager will make the right decision.

Finding an inventory manager you can believe in is of the greatest importance. Inventory control is gambling, risking the company resources on events determined by chance. A successful inventory manager is a professional gambler. He will make mistakes because he's gambling on an uncertain future.

The initial screening process is the one best tool for finding the right inventory manager. Here are three things to look for.

1. *Find a winner.* You need someone who's been a winner because he'll need self-confidence to handle the first big setbacks. He'll need self-confidence to sell himself and his ideas to your company.
2. *Find an expert.* You need someone who knows inventory control. There is a lot to know about inventory control. Even my little bookshelf at home has thirty-seven books on inventory control. You need a manager who knows what's available, someone who won't take twenty years to reinvent what's already been done.
3. *Make sure the manager is human.* Since an inventory manager

works through people, his whole job depends on his ability to win the cooperation of other people. Some inventory managers have spent so much time on theory that they have lost the ability to talk to people. Theories are great, but in the real world the real things that happen involve people.

What's a winner? Someone who made things happen—increased inventory turnover, improved customer service. You can tell by a person's record. The right person is on the move, growing, getting promoted, making more money. You'll probably pay top dollar for this person's services. If this person is the right one, his salary will be a drop in the bucket. If he's the wrong one, the losses from inventory excesses and poor customer service will overshadow any salary.

What's an expert? Not just someone more than fifty miles from home. The inventory control sciences have grown beyond common sense and hard work. Competition won't allow just anybody with intelligence to be an inventory manager. An expert spends years learning the job. Like any other profession, inventory control requires real formal training.

How do you know whether someone is an expert? Ask an expert. The recognized experts in inventory control are known and respected by their peers. They write books, lecture, teach, consult. They know the real experts from the term droppers. Ollie Wight, Joe Orlicky, Bob Brown, and Walt Goddard are experts, and there are many others. These people live and breathe inventory management. They are in daily contact with inventory control people. The experts' advice will cost you some money, but it will save you money and trouble in the long run.

What's a human? Determining that is probably the hardest, but you can still look at the record because a person probably couldn't become a winner unless he had a way with people. Sometimes technical ability alone can carry a person through the lower ranks of management, but it's not enough to reach upper management. People experts will tell you personality is the least reliable factor in testing, so for the human qualities you must rely on your own management experience.

Of course there are different styles of management. It is sometimes said that "good guys" make bad managers, but it doesn't necessarily follow that "bad guys" make good managers. An inventory manager doesn't have to be a social worker, but he must be human enough to understand what motivates people and human enough so people will understand what he's trying to do. Some funny things can happen when people don't understand what the inventory manager is trying to do. One brilliant inventory manager who used critical ratios for scheduling the plants had problems because the plant managers didn't understand the priorities and were afraid to ask how the system calculated the priority dates. One plant manager used to have his daughter put check marks on the report so the inventory manager would think he was using it.

You need more than an inventory manager who's a winner and an expert. You need an inventory manager who's human enough to see the real world.

Inside or Outside?

It's a tremendous advantage if the right inventory manager is already in the company. Don't lose someone you already have. Is he a winner? Compare his present performance against prior years. Is the customer service getting better? Is he reducing inventory in relation to sales? Has he reduced ordering and carrying costs? Measure his performance against prior years, not against some unrealistic goals the company has never achieved.

If you've got a winner, make sure you keep him. Is he an inventory expert? Has he kept up with the most recent developments in inventory management? If you don't know how to measure his expertise, bring in an expert in inventory management to evaluate your current inventory manager. Your inventory manager may at first resent this evaluation, but in the long run it's to his advantage. If you know your inventory manager is an expert, you can give your full support. He needs your complete confidence to improve inventory management performance.

Is he human? Inventory management involves all the company's departments. Do the buyers respect the inventory manager? Does he have a good rapport with the finance department? with the operations department? with the sales department? The inventory manager's job is getting things done through people. Without the support of the other operating departments in the company, it is impossible for the inventory manager to improve performance significantly.

Evaluating your current inventory manager is a difficult job. There's a lot of turnover and a lot of emotion in inventory management. Since the inventory manager's job is to gamble on an uncertain future, he's going to make mistakes. It's easy to say what his decision should have been after the sales results are in.

All inventory managers make wrong decisions. The right inventory manager just makes more right decisions than wrong ones. If you've already got the right inventory manager, make sure you don't lose him. But if you've got the wrong inventory manager, correct the problem quickly. The wrong inventory manager can cost the company thousands of dollars in lost customer service and inventory excess. The wrong person in inventory management will be miserable. Give him a chance to get into a job that matches his skills.

If you don't have an inventory manager or if you decide that you don't have the right one, you may still find the right inventory manager within the company. It's still a tremendous advantage to find someone within the company. You may have an outstanding systems person who's well versed in inventory management techniques. You may have a buyer whose skills go beyond the individual item. You may have a manager who has outstanding skills in getting things done through people. Whether you look inside or outside the company for the right inventory manager, nobody will be perfect. You're looking for the right blend of a winning record, an expertise in inventory management, and the ability to deal through people. You must have the right inventory manager before you can begin to create new systems.

Once you've found the right inventory manager, you must set realistic goals. The right inventory manager can set them for you.

SET REALISTIC INVENTORY GOALS

Now you have the right inventory manager. Here's how he sets realistic goals to improve customer service, to improve inventory turnover, and to reduce ordering costs.

We've seen how an inventory manager trades inventory investment for customer service and work reduction in the company. He does this by controlling safety stock time and review time. No matter how informal a company's inventory management system is, a relationship exists between inventory investment, customer service, and ordering costs. To set realistic goals, the inventory manager must first determine what this existing relationship is.

The things that determine the existing relationship include:

1. The accuracy of the existing forecasting system;
2. The mix of the items in the inventory, seasonal and nonseasonal, new and old, allocated and not allocated, low volume and high volume;
3. The size of the work force;
4. The availability of cash and space;
5. The performance of competition;
6. The quality of the buying decisions;
7. The gross profit rate;
8. The stated customer service and turnover goals;
9. The current economic trend; and,
10. The accuracy of the inventory records.

The inventory manager can get some understanding of this relationship by looking at the company's recent history of performance. He can project possible short-term goals for this relationship by simulating customer service results with varying review time and safety stock time decisions. Appendix 10 gives the details of this simulation process. Figure 11.1 shows an example of the results of the simulation process.

For last year's sales, the computer calculated the inventory turnover and customer service with different combinations of review and safety stock times. The inventory manager uses this relationship to set short-term inventory goals. Until the inventory

Figure 11.1 INVENTORY GOAL RELATIONSHIP

Simulation number	Review time (months)	Safety stock time (months)	Inventory turnover	Customer service (%)
1	1.0	0.0	22	73
2	1.0	0.5	10	84
3	1.0	1.0	7	88
4	1.0	1.5	5	91
5	1.0	2.0	4	93
6	1.0	2.5	3	94
7	1.0	3.0	2	95
8	1.0	3.5	1	96
9	0.5	0.0	45	82
10	0.5	0.5	15	87
11	0.5	1.0	8	91
12	0.5	1.5	6	94
13	0.5	2.0	4	96
14	0.5	2.5	3	97
15	0.5	3.0	2	97
16	0.5	3.5	1	98

manager can change this relationship within the company, these are the only realistic goals.

MORE ON SETTING REALISTIC GOALS

To: Inventory   1/9/this year
               manager
From: General manager
Subject: Inventory goals

I know you really haven't been with us very long, but I think it's important for you to understand our inventory goals early in the game.

We are a customer-oriented business dealing in fairly competitive products. If we don't have goods, our customers will buy from someone else. For that reason our customer service goal must be high. I realize that it is impossible to be in stock all the time. But I think it is reasonable to ask for and to demand a consistent 98 percent customer service level.

We are a growing company hard-pressed for cash. At this moment our single biggest problem is excess inventory. We buy every month. Theoretically we should have a turnover of at least twelve, but I understand that no system is perfect. Therefore, I want a consistent inventory turnover of at least six.

I've asked my administrative assistant to chart your progress toward these goals each month. Our current relationship between customer service and turnover is clearly unacceptable, so you have a fertile area for dramatic improvement.

It's already come to my attention that you are using our data processing resources for simulations. I urge you to focus full attention on our immediate problems.

To: General manager 1/13/this year
From: Inventory
           manager
Subject: Inventory goals

I don't really know whether we can ever get a 98 percent customer service level with six inventory turns. We certainly cannot with our existing inventory system and our existing product mix.

Quite seriously, the only time we accomplished these goals in the past ten years was the year our southern warehouse burned down during a mail strike.

Our inventory goals for one year from today should be a 91 percent customer service level with five inventory turns.

To: Inventory   1/20/this year
           manager
From: General manager
Subject: Inventory goals

Our goals for this year are a 91 percent customer service level with five inventory turns. These goals are an adequate improvement over our recent performance.

Our longer-range goals must be a 98 percent customer service level with six inventory turns. As you know, we are now revising our company five-year plan. Your plans should detail the steps necessary to meet our longer-range inventory goals. The plan should include (1.) year-by-year improvement in service and turnover, (2.) inventory dollar investment, (3.) systems and programming expense, and (4.) any added people or equipment expense in inventory control, merchandising, or warehouse operations.

We need the plan by March 31. I urge you not to neglect our immediate problems in service and turnover while you are preparing this plan. If we don't have immediate improvement, our five-year plan will be academic.

The final step in setting realistic inventory goals is confrontation. The company president knows what he wants the inventory goals to be. The inventory manager must know what goals are realistic. The inventory manager must set the short-term goals. Nothing will change the basic relationship of investment, customer service, and work in the company in the short term. The company president must set the long-range goals. The inventory manager must find ways to make these goals realistic and plan inventory objectives to meet those long-range goals.

Now we have the right inventory manager and realistic goals. Here's how to plan specific objectives to meet those goals.

PLANNING SPECIFIC OBJECTIVES

Many long-range plans are insults to the reader. "We will increase sales 42 percent. We will improve profits 53 percent. We are in business to service the customer. We will maintain excellent employee morale." But these long-range plans seem meaningless because they are so vague. Here's an approach that forces the manager to be objective and specific. Here's an approach that starts with a picture of the existing situation.

Things We Do Well Now

1. Inventory turnover has been five or better for the past year. We have no significant inventory excesses.
2. Customer service has improved 26 percent over the service levels during the energy crisis and recession years.
3. The purchasing agents change less than 8 percent of our computer's suggested buys.
4. For the first time in the history of the company, we have a professional expediting staff.
5. Money is available for inventory investment. For the past three months we have invested over $1 million in short-term high-interest paper.
6. The bulk of our inventory investment is in the best-selling items in the company. Our buying systems concentrate investment in our best-selling items.

7. We have computer programs that routinely get rid of excess inventories.
    a. Our computer prints work orders to move end-of-season excess to the warehouse branch that needs it.
    b. Our computer pinpoints excess quantities to use for our pool programs.
    c. Our computer picks very bad excesses for discount bulletins.
8. We plan inventory increases and decreases as part of the company budgeting process.
9. We have tight control over movement of priority merchandise within our company warehouses. We use a customer service key indicator reporting program.
10. When we have a customer service problem, we know why. We know the percentage of markouts caused by underbuying, vendor delivery, and inbound float delays.

Things We Can Do Better

1. We have four warehouses. Over 98 percent of the time customer markouts in one warehouse are available in another warehouse. We have an opportunity to improve customer service by shipping backorders from secondary warehouse locations.
2. With our existing computer forecasts, we could set up a distribution requirements planning system with our major suppliers. Distribution requirements planning would tell our key suppliers our expected requirements by item for the next six months. Our suppliers could use these plans to schedule production. This would cut their risk and improve our customer service.
3. Our buyers change few of the computer suggested buys. We could cut their work in half by separating items that require review—new items, big dollar buys, and items that show irrational demand patterns.
4. We have our own truck fleet. Our trucks backhaul a lot of our vendor shipments to us. We miss some attractive vendor volume discounts because we order for individual warehouses.

We could earn these discounts by picking up a consolidated order for our company and distributing the goods to the four warehouses.

5. Our buying systems concentrate our inventory investment in our best-selling items, improving our turnover and customer service level. We could show our customers how to use this same idea to improve their turnover and service. If all our customers carried all our best sellers, our sales would double.

6. Every month we make a list of our top twenty vendor delivery problems. About ten of our vendors are in the top twenty almost every month. We have an opportunity to increase our overall customer service level by concentrating our attention on these ten vendors.

Things We Do Poorly Now

1. We are too slow to react when the economic climate changes. We were well into the energy crisis before we started heavy buying. When the depression hit, we took three months to stop the building of inventory. It took another six months to reduce the inventory back to a reasonable level.

2. We don't trust our perpetual inventory balance. Every month we count almost 60 percent of our items. These counts can cause as many problems as they fix. Many "errors" are really just timing and cutoff problems. We must learn to trust our perpetual inventory.

3. We take too long to receive and put away priority orders. We never reach our goal of receiving and putting away all priority orders within three working days. Our warehouses feel that the existing paperwork systems make it impossible to receive and put away all priority orders within three working days.

4. We carry many lines for which the vendor minimum orders are very high. These vendor minimums include poundage minimums to make full truckloads, case minimums to make reasonable discounts, and ship pack minimums. Sometimes we must buy a year's supply or none at all.

5. Our truck systems are sometimes at odds with our customer service goals. Because we want to maximize backhaul revenue,

we sometimes ask a vendor to hold a shipment that we urgently need for customer service.

6. Our extremely conservative buying of high-price merchandise hurts customer service, especially at Christmas and shortly after.

7. We fail to get our new goods into our catalog at the same time that the goods come into our warehouse. This causes service problems when the new goods get in the catalog before they arrive in the warehouse and inventory excesses when the new goods arrive in the warehouse before they get in our catalog.

8. Our fixed vendor-review cycles make it hard to catch quick changes in seasonal sales trends. A severe winter can deplete our snow shovels in a week, and we may not buy again for a month.

Things That Could Hurt Us

1. If we accidently destroyed our computer sales records or our computer open-purchase order file, we wouldn't be able to buy right for six months to a year.

2. If there was a mail strike we couldn't send purchase orders to our vendors.

3. We have a lot on order. If the economy falls off, we could get buried with inventory.

4. A truck strike could delay most of the shipments to us and we would quickly run out of stock.

5. One year we had an energy crisis. Our vendors rely on energy and oil to make thousands of items for us. During the energy crisis we lost customers when we couldn't satisfy demand, and another energy crisis could happen.

The hard part of long-range planning is seeing our existing situation clearly, but other departments can help us. Ask the sales force what our faults are. They already have a whole list ready. The warehouses probably have another list, and the controller probably has still another. Most companies that use this approach form a group made up of the manager of each department in the company. Working together, they can give a beautiful picture of the company's existing situation. If the group identifies the objectives, they all work together better to achieve them.

Now that the inventory manager knows his existing situation, it's easy to come up with meaningful objectives. Just relate goals—better service, higher turnover, and less work—to the situation. Two rules for setting long-range objectives are forget how much it will cost and forget how long it will take. That comes later. For now, just figure out what the objectives should be and how to get there.

Goals

1. Improve customer service.
2. Increase inventory turnover.
3. Reduce ordering costs.

Things We Do Well Now

1. No excesses now
2. Improving customer service
3. Buyers trust the system
4. Professional expediters
5. Money available
6. Best-seller inventory
7. Automatic excess disposal
8. Budgeted inventory plans
9. Control of put-away
10. Reasons for markouts

Things We Can Do Better

1. Transfers between branches
2. Distribution requirements planning
3. Reduced buyer work
4. Consolidated orders
5. Best-seller distribution
6. Top twenty problem vendors

Things We Do Poorly Now

1. Slow to react
2. No trust in perpetual inventory
3. Slow receiving and put-away
4. Vendor minimums
5. Backhaul service problems
6. Conservative high-price buying
7. New goods coordination
8. Seasonal fixed-review buying

Things That Could Hurt Us

1. Loss of computer records
2. Mail strike
3. Downturn in economy
4. Truck strike
5. Energy crisis

Here's a sample long-range plan which shows how to combine a company's current status with its goals.

THE LONG-RANGE INVENTORY PLAN

Inventory Goals

Year	Inventory investment (millions of dollars)	Average turnover	Customer service level (%)
1	10	5.0	91
2	12	5.0	95
3	14	5.0	98
4	14	5.5	98
5	14	6.0	98

Here's how to achieve these goals.

I. Objectives and Procedures

Year 1

Objective: Safeguard existing customer service and turnover
1. Write procedures for our existing inventory control systems.
   a. Forecasting and buying formulas
   b. Receiving and put-away procedures
   c. Systems for phone and mail expediting
   d. Excess-inventory reduction programs
2. Set up controls to catch errors in demand, on-hand, and on-order records.
   a. Do today's customer orders add up to today's total demand by item?
   b. Do yesterday's on-hand totals plus receipts less shipments equal today's total on-hand?
   c. Do yesterday's on-order totals plus new purchase orders less receipts equal today's total on-order?
3. Set up disaster restart procedures to protect our computer master files.
   a. Once each month copy the demand, on-hand, and purchase order files.

    b. Store the copies in a remote location outside the main office.

    c. Store hard copies of customer demand, purchase orders, and receiving papers in a fireproof location.

    d. Twice a year copy computer inventory programs.

    e. Store the copies in a remote location outside the main office.

4. Make up a detailed inventory calendar for key inventory tasks.

January	Review spring and summer season prebuys.
February	Cancel overdue winter goods purchase orders.
March	Balance end-of-winter excesses by warehouse.
April	Set up more frequent reviews of spring and summer season items.
May	Make special buys to cover vendor shutdowns in July.
June	Release winter excesses against customer seasonal prebuys.
July	Cancel overdue summer goods purchase orders.
August	Run end-of-season discount bulletin on major excesses.
September	Review fall and winter season prebuys.
October	Balance end-of-summer excesses by warehouse.
November	Set up more frequent reviews of winter and gift season items.
December	Release summer excesses against customer seasonal prebuys.

Year 2

Objective: Improve customer service and keep our existing turn-over

1. Set up a distribution requirements planning system with major suppliers.

    a. Start with a local vendor who is a major markout problem.

    b. Get the vendor to agree to guarantee delivery in exchange for guaranteed purchase commitments.

    c. Set up a procedure to consolidate regular stock forecasts, pool orders, and promotion orders.

    d. Break scheduled net requirements into full-truck quantities.

e. Tie in vendor full-truck shipments with our own truck backhaul system.

f. Spread the system's use to as many of our major vendors as possible.

Eventually the planning and control of this system will replace the major part of our expediting work.

Year 3

Objective: Continue to improve customer service while keeping our existing turnover

Our strategy here will be to use other branch inventories to improve service. We can do this three ways:

1. Ship customer markouts in one branch from some other branch.

2. Backorder customer markouts for the length of time it would take to replenish the markouts from some other branch.

3. Anticipate customer markouts. Transfer goods from some other branch before the markouts occur.

We must analyze each alternative. Shipping markouts from some other branch increases freight costs. Who would pay the added freight cost? If we backorder customer markouts, we would complicate the customer ordering procedures and add to our order processing costs. We could anticipate customer markouts, but we may get the goods from some other branch just as our vendor's purchase order arrives at this branch.

Year 4

Objective: Improve inventory turnover and maintain existing customer service level

In branch warehouses with lower sales volume, our inventory turnover performance is lower, a direct result of increased buying to meet vendor shipping restrictions. The buys that we increase to make a truckload or get a quantity discount hurt turnover. We have an opportunity to improve turnover by consolidating our branch warehouse buying.

1. Find our major problems in meeting vendor minimum and quantity discount requirements.

2. Set up a system to summarize purchase orders for our branch warehouses for these vendors.
3. Set up a system to print automatic transfer papers when the goods arrive at the primary branch warehouse.
4. Balance the east/west and north/south vendors to minimize freight costs.

Year 5

Objective: Continue to improve inventory turnover and maintain existing customer service level

Putting our money in our best sellers is great for turnover. We rarely get stuck with excesses on best sellers. Now 20 percent of our items account for 80 percent of our business. We'd like 20 percent of our items to count for 90 percent of our business. Our strategy is to get our customers to concentrate on our best-selling items.
1. Give our customers a list of our best sellers.
2. Give our customers an incentive to buy the best sellers at least once.
3. Find out which customers aren't buying our best sellers.
4. Set up simple best-seller reorder systems.
5. Advertise and promote best sellers.

II. People Needed For This Plan

	Year				
Inventory control	1	2	3	4	5
Inventory manager	1	1	1	1	1
Clerical staff	2	2	2	2	2
Expediters	3	2	1	0	0
Inventory analyst	0	1	1	1	1
Distribution requirements planning staff	0	1	2	3	4
Total inventory control	6	7	7	7	8
Systems support	1	2	2	2	1
Warehouse	0	1	1	1	1
Total	7	10	10	10	10

III. Costs

Year	Strategy	Development	Operating
1	Control programs	$10,000	$5,000
2	Distribution requirements planning	50,000	25,000
3	Between-branch transfers	30,000	100,000
4	Consolidated ordering	30,000	50,000
5	Best-seller distribution	50,000	10,000
Total		$170,000	$190,000

IV. Benefits

Year	Strategy	Objective
1	Control programs	Protection
2	Distribution requirements planning	A four-point increase in customer service
3	Between-branch transfers	A three-point increase in customer service
4	Consolidated ordering	A 0.5 increase in inventory turnover plus added quantity discounts
5	Best-seller distribution	A 0.5 increase in inventory turnover plus added sales volume

In the plan year we will provide a detailed cost-benefit study of each objective and strategy.

SUMMARY

People, goals, and plans create inventory management systems that work. As we have seen, three steps are important in this process:

1. Find the right inventory manager.
2. Set realistic inventory goals.
3. Plan specific objectives.

Find an inventory manager who knows systems *and* people. Let the inventory manager set realistic short-term goals; let the company president set long-term goals. Get a picture of the company's current situation to plan specific objectives, and let the other managers in the company share in planning these specific objectives. Make sure the specific objectives meet the long-term goals.

Most of these inventory plans rely on computer systems. What kind of data processing management do we need to make these plans a reality? In the next chapter we will look at the role of data processing in creating new systems.

Chapter 12

What to Expect of Data Processing

The inventory manager and the data processing manager must work together to create computer inventory management systems that work.

In chapter 11 we saw that people, goals, and plans create inventory management systems that work. We saw that the right inventory manager sets realistic goals. And plans specific objectives. It is the data processing manager who must make these plans a reality. In this chapter we will look at people, goals, and plans in data processing management.

Data processing people have to become part of the company team. They must want to create inventory management systems that work. In some companies data processing is a line function with its own objectives. We must make data processing a real staff function. Data processing people must specialize in the department they support. To communicate with each other data processing people need to learn about business and line managers need to learn about data processing. Data processing people have to share in the risks and rewards of business, and we must evaluate them on how well their systems have achieved the objectives they were created for.

Company goals must become data processing goals, and data processing efficiency goals must be linked to company goals. Line managers must help set data processing goals, and goals must balance data processing efficiency and company effectiveness.

Data processing requests, like capital expenditure requests, require specific objectives. These plans must show how the objectives advance company goals. They must be specific and measurable.

In chapter 11 we saw how people, plans, and goals create the concepts for systems that work. In this chapter, we will see how people, plans, and goals make these concepts a reality.

PEOPLE

Data processing is a staff function. Thus data processing must support the line managers and they must be specialists in the departments they support. They must be part of the company team. Unfortunately, in many companies data processing people are more concerned with computer technology than with the business of the company. They must become part of the company team. They must want to create inventory management systems that work.

Some data processing specialists love the technical part of their job. They are fascinated with the speed and intricacy of computers. Some companies hire them because of their technical ability and fire them because they don't understand the company's business. Look at an ad in the newspaper for a data processing specialist. Does it say, Come to Giant Hardware because we pay high, because we're growing, because we're profitable, because we want you as part of the management team? No, it says, Come to Giant Hardware because we have the latest 380 system with on-line, real-time, virtual memory. Ask an accountant what he does for a living and he'll tell you he's chief accountant at Giant Hardware. If you ask a programmer the same question and he tells you he's a programmer on the 380 system with on-line, real-time, interactive virtual memory, that programmer's future is his technical knowledge, not the company.

Some of our colleges are guilty of the same emphasis on technical rather than business expertise. I interviewed a man with an associate degree in data processing who knew the history of data processing from punched cards and Eniac through the Sabre System. He knew FORTRAN, BAL, and COBOL. He could even multiply in binary. He knew how the electricity flowed through the little doughnuts inside the computer core memory. But he didn't know about profitable information systems.

There are ways to make data processing people more a part of the company. In January 1966 when I became the director of information systems for a $40 million division of our corporation the computer section cost $1.3 million a year, more than 3 percent of sales. Managers complained they couldn't do their jobs without data processing support. Systems were always late, and managers didn't believe the reports when they got them. (The reports *were* wrong.) The computer section was completely demoralized.

My boss wanted to know how long it would take to clean up the backlog of systems requests. There were 245 systems requests. Without any new computer requests we optimistically estimated completion in 16.6 worker years. There were sales analysis requests that would restructure the entire company's market direction. There were inventory control requests that affected customer service, company return on investment, and all the people working in the company's fourteen plants. There were financial requests that would eventually change the company's gross profit and expense structure.

My problem was to organize the department for action. We had the most friction in inventory and production control. The vice-president in charge was brilliant and driving and he had an inventory manager who was also brilliant and driving. When the inventory control manager was fired, he claimed that lack of data processing support had been his downfall. He would have liked nothing better than to prove he was right, so I hired the ex–inventory control manager as an inventory control specialist. He and his project team were to handle all data processing requests from the inventory and production control vice-president.

It worked! The vice-president finally had somebody in data processing who talked his language, somebody he knew would be working on his requests day after day. It worked so well that the other vice-presidents wanted specialists for their areas. Eventually we had four specialists—one in inventory and production control, one in marketing, one in accounting, and one in general administrative.

Only the inventory and production control specialist was hired from another department. The other specialists were previously programmers. It didn't seem to matter. The accounting specialist who had been a programmer soon identified with the accounting department. He went to the National Accountant Association meetings, read all the latest releases from the Accounting Principles Board, studied accounting at night. He was the controller of one of the international divisions the last I knew of him. Data processing was becoming part of the company.

BRIDGING THE TECHNICAL GAPS

One way to make data processing people part of the company is to have them specialize in the department they support. The next step is to teach the data processing people about business and to teach the line managers about data processing. But there are real language problems when data processing people and line managers talk to each other.

Line managers say: What's the ROI on this capital expenditure? What's the payback period? What's the market potential? How's our asset turnover? Here's an example of the KITA management style. Let's redo our five-year plan. We use flexible budgets. What's the best media mix? We must level production. Our customer service stinks. Our bad-debt reserve is inadequate. The strategy is planned obsolescence. How's our worker morale? What's our inventory turnover? Our lead times are reasonable. We need a gross profit mix analysis. Let's develop a consumer franchise. Here's a value analysis of our product. In this design year we will retool. The indirect expense is high. Let's encourage upward communications.

Data processing people say: This is a real-time system. Our information system depends on direct access. We use CICS. The interrecord gaps are destroying throughput. We have a basic BOMPs supporting our MRP. We use POWER. BAL is more efficient than COBOL. Our data base has a high degree of redundancy. The hex dump pinpoints data exceptions. We have to reprogram to provide restart checkpoints. The grandfather, father, son procedure is ineffective in real time. Multiprogramming reduces wait time. Our communication system depends on transparency.

There are many ways to teach data processing people about business. One is the direct approach. Ask line managers to teach them. This has some side advantages. Many older executives who came up the hard way with no formal education view programmers as threats. These executives don't want to ask questions that may expose their ignorance about computers. When they start to teach data processing people, they find out just how ignorant the programmers can be about business. Lawrence Appley once said, "Knowledge is the supreme authority." Teaching reinforces the executive's authority image. When the executive understands just how valuable his own experience is, he is more willing to expose his ignorance about computers. So when the executive teaches data processing people about business, he almost automatically learns about data processing.

There are many ways to teach line managers about data processing. There are many user information retrieval packages that let line managers generate reports from the computer. These report generators are crude and inefficient, but they make the line manager aware of the frustrations that programmers experience every day. Programs have thousands of detailed instructions, and the computer does not tolerate any deviations from its rigid logic. Sometimes a line manager doesn't get a report from the computer because he left out a comma or a dollar sign. He starts to understand the rigors of that discipline. At the same time the line manager quickly learns to think through what he really wants from the computer the first time. Fuzzy thinking is just too time-consuming. User information retrieval packages are great training aids for line managers.

The most powerful tool for teaching line managers and data processing people I've ever seen is video tape. We used Ollie Wight's tape on inventory management to spark a whole new inventory and production control system. Together the company inventory control executives and the data processing people watched Ollie use the new computer tools on the age-old problem of inventory management. The tape itself was full of new ideas, and it was easy to understand complicated topics explored in a conversational mode.

The most exciting part was the dialogue between our data processing people and our line managers. "Do we have that problem in Dothan, Alabama?" "No, we don't have that problem.

We're so far behind we won't have that problem for another three years. Here's the problem we have right now." "Could we do that? I never imagined the computer could schedule a plant let alone schedule the raw materials."

These video tape sessions start with everyone listening and watching the tape. They end with as much dialogue between the viewers as there was on the tape. The right video tape is worth a thousand times its cost. It's the cheapest way to teach data processing people and line managers to understand each other.

Teaching the company president about data processing is important, because it teaches him how to measure performance. But most computer schools teach hardware and software rather than the profitable use of data processing resources. When our company president went to computer school, he learned all about tapes, disks, core storage, and programming. He even wrote a program that simulated the path of a bouncing rubber ball. When he came back he sent his line managers to the same kind of school, so they too learned about tapes, disks, core storage, and programming. They could ask, "Are we using a data base?" or worse, "What operating system do we use?" They never learned what makes an information system profitable. But it was a start.

The first step in making our data processing people part of the company is to have them specialize in the department they support. The next step is to teach the data processing people about business and to teach the line managers about data processing. The final step is to let the data processing people share in the rewards and risks of the systems they develop.

SHARING REWARDS

Data processing people probably work more overtime than any others in the company. For the most part they are bright and hardworking. Their work demands rigorous logic, but their efforts are rarely measured fairly by the other departments in the company.

Sharing rewards doesn't just mean paying high wages. It means taking the time to explain why something is needed. It means asking for help before decisions are made. It means letting the data processing people share the risks. Business is risk-taking for a profit.

If you let the data processing staff share the thrill of that risk-taking you've made them part of your company. If you make them robots who just carry your instructions to the computer, you've created career computer personnel whose loyalties are to the technology. If you make them part of your company team, they will want to create inventory management systems that work.

To evaluate systems, you can't just measure how much the system costs to install and operate. You have to measure how well the system achieves its goal. Measure the data processing people on effectiveness, not just efficiency, and let them share the risks and rewards.

Letting the data processing staff share the rewards doesn't necessarily mean paying them incentive compensation on the success of their systems. It means inviting them to meetings that pertain to improving customer service and inventory turnover. It means letting them talk to top management when the discussion concerns inventory management. It means giving them copies of the reports on the system's performance. It means asking their advice about a problem before deciding how to solve it. It means letting them share in the criticism when the system doesn't accomplish its objective.

When the data processing staff share risks and rewards, they become part of the company. When they are part of the company, they help create systems that work.

Now let's look at goals in data processing.

GOALS

Systems that achieve company goals are systems that work. The goals in data processing should be the same as the goals in the line departments. Data processing has efficiency goals of its own, but these must be secondary to the goals of the line managers. The goals in many data processing departments are not the company goals.

In some companies if no new systems were created, the data processing staff would remain just as busy, improving the efficiency of what they are already doing. But the way to create computer systems that work is to subordinate efficiency goals to the goals of the company's line managers.

I surveyed fourteen companies whose sales ranged from $14 million to over $1 billion. They expended over 75 percent of their data processing effort to improve what they were already doing. Convert to a new computer, to a new operating system, to a new programming language. Streamline the order entry system. Speed up the payroll, speed up the accounts receivable. Modify the general ledger package. Build a better data base. Redo the documentation.

They were asking all the wrong questions. What computer should we use? How long will it take? How much will it cost? Should we convert to OS? What percent of sales do data processing costs represent? Should we centralize or decentralize? What are the line costs? Should we process in batch mode or use direct access? How's our documentation? How many key strokes per hour? How many clerks can we save? What language should we use—COBOL, BAL, or RPG? Do we have a data base? Should we charge out computer time? Are our programs efficient? Are we multiprogramming? To whom should data processing report? Should we buy the computer or rent it? Why does it take so long? What's our backlog of work in worker years?

These are secondary considerations. The answers are always wrong because the questions are wrong. Data processing efficiency goals should be part of the company line manager's business goals.

It is the data processing manager's job to use data processing resources to meet company goals. From all the requests for data processing support, the manager must single out those that best support company goals, and he must balance internal efficiency goals with company goals.

When I became director of information systems in 1966 I found requests such as this in the previous data processing manager's file:

```
To:    DP manager
From: Chief purchasing agent
Please add the letter J to our list of
```

available buyer codes as we have hired a
new employee, Alex Jones.

To: Chief purchasing agent
From: DP manager
We cannot add the letter J for Alex Jones.
Our letter system is set up for the major
product divisions of our company. A
through I are durable purchased products.
J through R are manufactured domestic
products. S through Z represent your
product line. The only letter possible for
Jones is Q. But you must give up the
international breakdown since you are now
using Q to break down Harry Brown's
domestic and international purchases.
I recommend that you wait until we have the
facility for a double-digit coding
structure. We are in the process of
revamping your entire product file. Our
next scheduled meeting with you is in
October. Can we discuss it at that time?

To: DP manager
From: Chief purchasing agent
I need to measure Alex Jones right away. I
can't give up the letter Q. Can't you give
me a ☐ or something for Jones right away?

To: Chief purchasing agent
From: DP manager
Sorry I'm so late responding to your memo
on Jones' letter code. As you know JGB has
imposed a moratorium on program changes
for the next six months. Even so we
analyzed your request.
    There are twenty-three programs that
edit buyer codes. We estimate to change

```
these programs to accept nonalphabetic
codes will take 2.3 worker months. If you
want to talk to JGB to extend our six-month
moratorium to handle your request I am
willing.
```

My first step was to form a management steering committee to put priorities on the existing requests and select requests that supported company goals. As we worked on each line manager's requests we streamlined the systems that supported them. The company goals became our data processing goals, and these goals helped us create systems that worked.

PLANS

The management steering committee that selected data processing priorities was made up of all the line managers in the company, the same people who made the 245 systems requests.

New data processing systems are very like capital expenditure programs. They involve large sums of money. They involve risk. Sometimes they work, sometimes they don't. They are long-term projects. Each project should be measured on its own cost–benefit ratio, yet each project should be part of a long-range plan for building a total data processing system. For every system request data processing must have a plan that shows the objective of the system and how it supports a company goal. The objective should be specific and measurable.

To make sure that every new systems request was part of a plan with specific objectives, we designed a systems service request form for all new requests. The line managers used this form for all new systems requests, and we asked them to fill out the form for the existing 245 requests. We insisted that the line manager put the requests in order of priority. Figure 12.1 shows this systems service request form.

The line manager filled in the first two questions. For example:

What: Install a distribution requirements plan with our major suppliers. We want to be able to give our suppliers a firm purchase order commitment some months into the future. We want to give

Figure 12.1   SYSTEMS SERVICE REQUEST

Number_____

Line manager_____   Date_____

1. What   (Tell us in general terms what you would like us to do)

2. Why   (Describe in measurable terms how this request will improve profits, increase customer service, or otherwise help our company)

Data processing_____   Date_____

3. How   (This is the way we plan to do your request)

4. How much to develop   (This is our development cost)

5. How much to operate   (This is the system's operating cost)

---

them a plan of our purchase requirements six months into the future. See year 2 in our company five-year plan for details.

Why: Right now we surprise our major suppliers with a purchase order every month. Our major suppliers may or may not have goods available to ship us. We want to trade firm future purchase orders and plan of our purchase requirements six months into the future for a promise of guaranteed delivery and consistent lead time. We can measure the success of this system by reporting the current months on hand and line-fill rate versus the future months on hand and line-fill rate for the following vendors: Black & Decker, Corning, True Temper.

The data processing staff filled in the "how" and "how much."

How: We will give you the ability to change the average lead time to a vendor promised lead time. This will allow you to generate firm purchase order commitments for future delivery.

We will change the forecast simulation program to project demand six months into the future and print this demand in the distribution requirements plan format.

How much to develop:

Systems development	Worker days	Dollar cost
Research, report design, flow charts, forms, documentation, meetings	300	$30,000
Programming		
File maintenance, program changes, operating systems	200	$20,000
Total cost	500	$50,000

How much to operate: Cost for 200 vendors on the distribution requirements planning system:

Computer time, clerical staff,
forms, tapes, disks, etc. $25,000

When we asked the line managers to use this form for the existing 245 systems requests, we found that nobody claimed 27 of the 245 requests and the data processing staff had made 48 of the requests. Of the remaining 170 requests over half were discarded when the department line manager tried to answer the question why. The 80 requests that then remained were changed. When the line manager saw the question asking *why* his department wanted something, he found out *what* they really wanted.

The steering committee reviewed all the systems service requests to set total company priorities for data processing. This approach gave us specific objectives that moved our company toward its goals. We streamlined our own data processing efficiency in the process of creating each new system.

CREATING SYSTEMS THAT WORK

The secrets to creating inventory management systems that work are people, goals, and plans.

In chapter 11 we saw how to find the right inventory manager, one who is an expert not only in inventory management but also in people. We saw how the inventory manager set short-term inventory goals and specified objectives that meet the president's long-range goals.

In this chapter we saw that the data processing staff must become part of the company team by specializing in the department they support. We've seen that education is the way to foster communication between line managers and the data processing staff and that the best way to make data processing part of the company team is to let them share the rewards and risks of the systems they help create.

Data processing departments have their own efficiency goals, but they must accept the company goals as their own. We've seen that a steering committee of line managers can select data processing goals and insure that data processing goals are company goals.

Systems requests, like capital expenditure requests, should explain what is to be done and why, how it will be done and how much it will cost. Systems requests are plans with specific measurable objectives that support company goals.

The inventory manager and the data processing manager must work together to create successful systems. People, goals, and plans create inventory management systems that work.

THE FUTURE OF INVENTORY MANAGEMENT

This book is about simple systems, built on the inventory management contributions of many people over the years.

I'm sure that even as you read this, other people are developing new systems. Inventory management is a classic business problem, and business is changing every day. Today's systems will not answer tomorrow's inventory problems. It's up to you to develop simple systems of your own. It's up to you to plan for tomorrow's inventory problems. We don't know what those plans will be, but as Larry Zehfuss, finance manager of American Hardware Supply Company said, "If we don't plan for the future, we won't have a future."

I hope this simple book has given you something of value for your future inventory management plans.

Appendix 1

THE REPLACE SALES SYSTEM

The finished goods inventory management systems in this book require the computer to record on-hand and on-order for each item. It is practical to record on-hand and on-order in manufacturing and wholesaling companies. It is even practical to record on-hand and on-order in retail distribution centers. But it is very expensive to record these figures in a retail store. Retail sales volume per item is too small to pay the expense involved in recording these figures for an item. Yet retailers must reorder goods just as manufacturers and wholesalers must reorder goods. This system describes a method of reordering whenever it is not practical to record on-hand and on-order.

Retailers have hundreds of thousands of items with relatively low-volume sales. No matter how clever a *retail* inventory control system is, its first measure of effectiveness must be its cost to operate per item. The only system that has proven effective for retail inventory is J. C. Penney's Semiautomatic Stock Control System (SASC). In 1965 when J. C. Penney's volume was $2.3 billion, half their unit sales in 1700 stores were automatically replenished using SASC. SASC cost less than half a cent per unit to operate. (Adding in store costs, it's still less than 1 cent per unit to operate.)

The beauty of SASC is what the computer does *not* need to know. The computer does *not* need to know the on-hand balance in the store, the model stock, the seasonality curve, the on-order for the item, store transfers, customer returns, or vendor lead times. All the computer needs to know are current sales and negative orders. Here is how SASC works:

1. At the beginning of a retail quarter the store department manager makes one physical count of the department's inventory by item.

   Item 23456   On-hand 40 pieces

   This on-hand balance is *not* put in the computer.

2. At the beginning of a retail quarter the department manager creates a model stock for each item in the department. The model stock reflects the number of units the department manager wants to have on hand at any time during the season.

   Item 23456   Model stock 50 pieces

   The model stock is *not* put in the computer.

3. At the beginning of a retail quarter the department manager writes one purchase order to bring the item on hand up to the model stock.

   Item 23456   Purchase order 10 pieces

   The manual purchase order is *not* put in the computer.

4. From then on, as frequently as once a week, the computer replaces sales.

   Item 23456
   This week's sales 12 pieces
   This week's purchase order 12 pieces

If the department manager wants to increase the model stock, he just writes a manual purchase order for that quantity.

Item 23456   Purchase order 10 pieces

The manual purchase order raises the model stock by increasing the stock on hand in the store. The manual purchase order does *not* go into the computer.

Item 23456   Original model stock   50 pieces
              + Manual purchase order   10 pieces
              = New model stock   60 pieces

If the department manager wants to decrease the model stock, he writes a negative order. The negative order *does* go into the computer. The computer does not replace sales for the item until the negative order quantity is reduced to zero by subsequent sales.

Item 23456   Negative order   −20 pieces

The computer would not replace sales for the next 20 pieces of sales. So the model stock would be reduced from 60 pieces to 40 pieces.

The strength of this system is its simplicity and low cost. It is probably the only system inexpensive enough to work in a retail store on a continuing basis. Perpetual inventory systems rarely work, and cost more to operate than the profit on low-volume items. The first measure of any retail inventory control system must be its cost to operate per item.

Appendix 2

PROFILE FORECASTING

Focus forecasting is a method of forecasting demand for items that a company offers for sale on a continuing basis. Companies offer some items for sale only once as part of a promotion, a fashion line, or a special event. Here the company only gets one chance to forecast the expected demand. Profile forecasting is a procedure that uses an early sampling of key customer demand to forecast the total demand expected for an item.

Every two years millions of Americans watch a very accurate forecasting process on television. Based on early election returns, networks predict election winners with as little as 4 percent of the vote counted. These predictions have proved so accurate that candidates will concede elections with less than half the votes counted.

In 1966 we used profile forecasting to project sales of a shirt producer. Just as 20 percent of a company's items account for 80 percent of its sales, 20 percent of a company's customers usually account for 80 percent of its sales. We studied three years' sales to find the percentage contribution of the major customers in a shirt category (fig. A2.1).

We treated these customers like key counties in a television election forecast. As demand arrived for dress shirts, we used these key customers to project total demand for that shirt (fig. A2.2).

This projection was particularly valuable in deciding whether to run a second round of production for tailor-made shirts. Since the

Figure A2.1 DRESS SHIRT DEMAND

Customer name	Customer dollar demand	Percentage of dollar demand	Cumulative percentage of dollar demand
Filene's	$52,500	1.64	1.64
A&S	46,800	1.46	3.10
Broadstreet	39,300	1.23	4.33
Dunhams	38,700	1.21	5.54

Figure A2.2 SPRING DEMAND PROJECTION

Item 70986		Pattern 1359B
Key customer demand	Percentage of key customer demands complete	Total current demand
$89,434	48.74	$107,842
Projected spring demand		183,492
Balance of expected demand		75,650

fabric was available, the question was whether to make more of a nonstock shirt like 70986 or whether to use the fabric to make a lower-cost shirt for stock.

Projected spring demand was simply the total current demand from key customers divided by the cumulative percentage of prior years' total spring demand represented by these key customers.

$$\text{Projected spring demand} = \frac{\text{Key customers current spring demand}}{\text{Percentage of key customers demand complete}}$$

$$\$183,492 = \frac{\$89,434}{0.4874}$$

In practice, profile forecasting had one serious limitation. Sometimes new customers would seriously distort the overall key customer percentage of total current demand. In those cases the total current demand in the period could exceed the projected demand for a short time. Nothing destroys credibility more than a forecasting system that predicts fewer sales than have already occurred. In any event, profile forecasting did provide useful direction to the shirt production decisions.

Appendix 3

A RANKING SYSTEM

Focus forecasting is used for items that a company offers for sale on a continuing basis. Appendix 2 described profile forecasting used for one-time offerings where early demand indications are available. The ranking system in this appendix is also used for one-time offerings, since it can project item demand without any early indication of item demand.

Practically everybody knows the 80/20 rule: About 20 percent of the items in inventory generate 80 percent of the sales. There is only one system I know of that uses the 80/20 rule to forecast sales.

Usually a lingerie company offers three separate seasonal new lines—a spring line, a fall line, and a holiday line. Lingerie designers create as many as 400 items to offer in a spring line. The future of the company and even more certainly the future of the company president lie in their ability to plan inventories to support the coming season. For this reason the most important talent of these chief executives is taste. Forecasting ability in lingerie companies is a measure of taste—not necessarily good taste. It's just necessary that the taste be consistent with that of potential customers.

In 1965 Dupont, which supplies raw materials to lingerie companies, developed a new forecasting system for new-line offerings. The forecasting system used the 80/20 rule. Here's how a lingerie company used the system to forecast the spring line. First the system would calculate demand curves for the last three spring lines. The system would rank each of the 400 items in each year from the best-selling item to the worst-selling item in terms of dollars of demand. The 1964 spring line consisted of 400 items; the total dollar demand for that spring line was $3.5 million. The curve analysis is shown in figure A3.1.

Figure A3.1   CURVE ANALYSIS, SPRING LINE 1964

Item	Item dollar demand	Number of items	Total dollar demand	Percentage of line items	Percentage of total dollar demand
23457	$205,100	1	$ 205,100	0.25	5.86
10013	196,250	2	401,350	0.50	11.47
34968	144,195	3	545,545	0.75	15.59
24074	126,505	4	672,050	1.00	19.20
15693	119,250	5	791,300	1.25	22.61
00904	118,550	6	909,850	1.50	26.00
10787	115,200	7	1,025,050	1.75	29.29

The system calculated this same analysis for 1965 and 1966. The graphs of the three years of spring lines are shown in figure A3.2.

So in three years 20 percent of the items represented between 62 percent and 82 percent of the total dollar demand. But the lingerie inventory planner projected a much less radical forecast almost every time. Could he risk gambling as much as 30 percent of the total inventory investment on 8 of the 400 items? Could he risk almost no safety stock on 100 of the 400 items? Hardly ever, and so the lingerie company had consistent shortages and heavy over-stocks and markdown losses.

The Dupont system could calculate the probable curve for 1967 based on the previous three years' history, and it could calculate the confidence limits for projecting the curve. The next step was to rank the 1967 spring line offering from the best seller to the worst seller.

To do this, the Dupont system used a technique similar to group consensus. When I attended a seminar on group dynamics, I was startled to learn the effectiveness of a group consensus. Each of ten persons in our group read a research paper on a management technique and took a multiple choice test on it. The highest individual score was 84 percent correct. Before our individual tests were graded we met as a group and picked answers to the multiple choice test through a process called consensus. The group analyzed each question through discussion. A simple majority vote was not enough to determine the answer to a question. Each member of the group had to be convinced that the answer selected was or at least could be the correct answer. In the time limit our group was only able to answer 96 percent of the questions, but *all* were correct!

The Dupont system used a similar technique to rank the spring line. All the managers met in the board room to rank the spring line. (Later it was decided to leave out the chief executive since he exerted so much influence on the group decision.)

The last step in the ranking system was to estimate the spring line's total sales. It is always easier to estimate total sales than to

Figure A3.2   SPRING LINE CURVE ANALYSIS GRAPHS FOR 1964, 1965, 1966

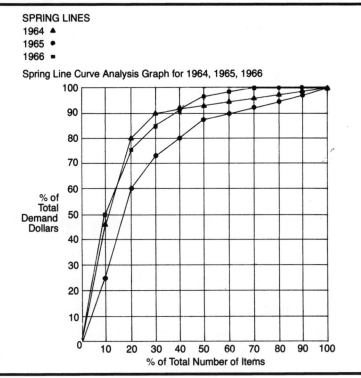

SPRING LINES
1964 ▲
1965 ●
1966 ■

Spring Line Curve Analysis Graph for 1964, 1965, 1966

estimate sales by individual item. Once the total dollar sales were estimated the previous year's curves were used to project sales for the spring line of 1967.

It is hard to judge just how successful the ranking system was. Projections for many of the items were so daring that the inventory planner would not make the gamble. Overall, though, it was a good year for the lingerie company. Deliveries were excellent and markdowns were lower than during the previous three years.

This forecasting system is the only one I've seen that makes use of the 80/20 rule. It is the only system I've seen that applies a systematic routine to forecasting one-time sales, and it produces usable results.

Appendix 4

ECONOMIC MODELS

Focus forecasting only uses *internal* demand history to forecast future demand. Some companies use economic models to forecast future demand. These models are usually expensive to create and costly to maintain. For these reasons companies usually use them to forecast total company sales, complete product category sales, or individual item sales in the millions of dollars. This appendix does not describe an economic forecasting model, but it will give you an insight into the workings of economic models in general.

Even the largest company is a chip floating on the turbulent ocean of the national economy. The results of business decisions depend on competitors' decisions. For the past two years I've been a board member of the Management Decision Games at Carnegie Mellon University. These games teach budding executives a basic rule of business life: "Results are the only measure of business decisions." There is fierce competition in business, but there is no par. All that counts is whether your company played the game better than the competition.

To allocate sales to the competing companies, the Carnegie Mellon computer uses an economic model including the following rules.

1. An overall economic trend influences all the company sales in the game "world."
2. The relative prices of each company's products influence each

company's market share in the game world. The lower a company's prices are the higher that company's sales are.

3.   The more a company spends for research and development now the higher that company's sales will be in future years.
4.   The more a company spends for advertising on a consistent basis the higher that company's sales will be in future years.
5.   The greater a company's ability to fill orders from stock, the more sales it gets now and in future years.

The economic model rewards consistent marketing strategies. Cutting prices for one year, or overspending for research and development, or saturating with advertising generate disproportionate sales increases. Companies with inconsistent marketing strategies lose money. This game teaches the rewards of sound management. It teaches the students to think in terms of results rather than techniques. The game itself demonstrates the economic model technique of forecasting.

Figure A4.1 is a matrix showing some different ingredients in this economic model. The computer calculates the expected sales for each company using some predetermined weighting for price, research and development, advertising, and customer service. In the real world, the problem becomes twofold: (1) How do you know what your competition is doing? (2) How do you know the real weighting that the consumer attaches to each decision? The matrix analyzer in appendix 5 shows a way to answer these questions.

Figure A4.1   TOTAL WORLD SALES TREND INDEX 1.05

Company	Pricing decision	Research and development	Advertising expenditure	Customer service (%)
A	$3.45	$40,000	$180,000	87
B	3.20	0	200,000	70
C	3.50	40,000	300,000	91

Economic model forecasting systems force a company to think through its place in the economy, a healthy exercise. Economic models rarely work well over long periods of time because business is dynamic while economic models, which are structured and complicated, quickly lose their forecasting effectiveness without continued maintenance.

Appendix 5

THE MATRIX ANALYZER

Appendix 4 described the general workings of an economic model. The first step in using an economic model to forecast is in itself useful and practical. An apparel company I worked for was a pioneer in consumer market research. Our market research director hated to make decisions without facts. Under his direction our company joined five other companies to ...iance a consumer research panel made up of 7000 families who formed a cross section of the United States. They recorded their daily family purchases in a diary.

Our company developed a unique information retrieval system called a matrix analyzer to summarize the data. Since we later sold the program to other members of the group for over $35,000, I can describe this system in only general terms even now.

The key to using the matrix analyzer was to ask the right questions. The program cross-referenced characteristics of consumers with characteristics of products. The form for asking a question looked something like the one in figure A5.1. The question was, "What is the percentage distribution of bra purchases by brand and age of female?" The resulting report looked like the one in figure A5.2. The answer to this question directed our designers to make bras for the mature woman.

For a forecasting viewpoint the matrix analyzer gave guidelines for new product sizes and color distributions. It provided the input for a usable economic model. We never programmed the economic

Figure A5.1 MATRIX ANALYZER

Row	Column
Price point	Age
Fabric	Income
Style	Education
Product group	Family size
Color	Geographic quadrant
Size	Sex
Branch	Store type

Raw units
Dollars
Percentage distribution
Projected units

Figure A5.2 BRA PURCHASES

	Age						
	0–19	20–25	26–30	31–35	36–40	41–50	Over 50
Playtex							
Maidenform							
Warners							
Bali							
Gosserette							
Exquisite Form							

model into the computer. But once we knew the market segment for a new product, we were more accurate at projecting the initial demand potential.

Appendix 6

STATISTICS VERSUS SIMULATIONS

There are two ways to predict what will happen in the future. One is based on probability; the other is based on simulation. What follows is the classic dice game example.

Probabilities are odds. In shooting dice, the odds are that the shooter will win on the first throw twice as often as he loses on the first throw.

The shooter plays by rolling two dice numbered 1 to 6. If the dice add up to 7 or 11 on the first throw, the shooter wins. If the dice add up to 2, 3, or 12 on the first throw the shooter loses. Figure A6.1

Figure A6.1   POSSIBLE TOTALS OF DIE A + DIE B

A/B	1	2	3	4	5	6	
1	2L	3L	4	5	6	7W	W = win
2	3L	4	5	6	7W	8	L = lose
3	4	5	6	7W	8	9	
4	5	6	7W	8	9	10	
5	6	7W	8	9	10	11W	
6	7W	8	9	10	11W	12L	

shows how to calculate the odds of winning or losing on the first throw.

Since there are 36 possible combinations, we can figure out the probabilities for the totals by dividing by 36 the number of times that total appears. Thus:

The winning dice throws are

Total (A + B)	7	11
Probability	6/36	+ 2/36 or 8/36

and losing dice throws are

Total (A + B)	2	3	12
Probability	1/36	+ 2/36	+ 1/36 or 4/36

The rest of the possible throws are
Total (A + B) 4, 5, 6, 8, 9, 10
Probability 3/36 + 4/36 + 5/36 + 5/36 + 4/36 + 3/36 or 24/36

So, on the first throw the probability of winning is 8/36, of losing is 4/36, of trying to make a 4, 5, 6, 8, 9, or 10 on the next throw is 24/36.

So, on the first throw the probability of winning is 8/36, of losing is 4/36, of trying to make a 4, 5, 6, 8, 9, or 10 on the next throw is 24/36.

So the dice shooter rolls again 24 out of 36 times, 67 percent of the time; of the other 33 percent of the time he wins twice as often as he loses. This is how the probabilities or odds are calculated.

A person can calculate the same odds by using simulation, that is, rolling the dice and keeping a record of what happens—how many times he wins on the first throw, how many times he loses. The trouble is that the only way he can come close to exact probabilities is to throw the dice all night long. But a computer can theoretically throw the dice a thousand times a second, so it can simulate almost the exact odds. Here's how the computer simulates the odds in this example.

Here's a list of four-digit numbers from a page in the telephone book:

0220	3460	6681	3171	9068
1373	2979	3543	7223	7272
9465	9523	5256	0801	8064

Adding the four digits in each number together generates a table of random numbers from 1 to 36:

4	13	21	12	23
14	27	15	14	18
24	19	18	9	18

If the computer is programmed to lose whenever it gets a number from 1 to 4 and win whenever it gets a number from 5 to 12, it can simulate rolling the dice. The chance of getting the numbers 1 to 4 is the same as the chance of rolling a 2, 3, or 12 with real dice, and the chance of getting the numbers 5 to 12 is the same as the chance of rolling a 7 or 11. Figure A6.2 shows the logic of programming a computer to do this.

Figure A6.2 FLOWCHART SHOWING SIMULATED DICE SHOOTING

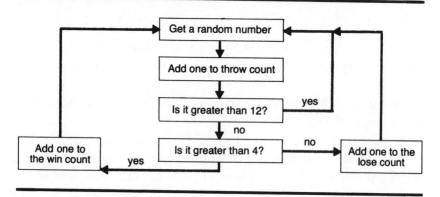

Let's say the computer has rolled the dice fifteen times and gets the results shown in figure A6.3. After fifteen throws the computer calculates that 13/15 or 87 percent of the time the shooter rolls again. The other 13 percent of the time he wins twice as many times as he loses. Now the computer doesn't roll the dice just fifteen times. The computer rolls the dice millions of times. Eventually it comes up with the same odds as the probability calculation.

So why bother simulating when probabilities are so much easier to calculate? Well, that's true with dice. But in inventory control you don't know all the variables affecting demand. In inventory control you must calculate the odds without knowing how many sides the

Figure A6.3   SIMULATED RESULTS OF FIFTEEN DICE ROLLS

Random number	Wins	Loses	Throws
4		1	1
13			2
21			3
12	1		4
23			5
14			6
27			7
15			8
14			9
18			10
24			11
19			12
18			13
9	1		14
18			15
Total	2	1	

dice have. And in inventory control what look like random occurrences are sometimes not. Statistics provide rules for gambling when the variables are known. Simulation provides rules for gambling when the variables are unknown. In inventory management many of the variables are unknown. Simulation approaches provide better rules for gambling than statistics do in gambling inventory dollars for customer service.

Appendix 7

EXPONENTIAL SMOOTHING VERSUS FOCUS FORECASTING FOR THE BROILER PAN

Let's see how exponential smoothing would forecast broiler pan demand for August, September, and October. We have the demand shown in figure A7.1. This is how exponential smoothing works.

Figure A7.1 A BROILER PAN DEMAND IN UNITS

	Jan	Feb	Mar	Apr	May	June	July	Aug	Sept	Oct	Nov	Dec
Last year	6	212	378	129	163	96	167	159	201	153	76	30
This year	72	90	108	134	92	137	120	?	?	?		

1. Select a moving average period. Some exponential smoothing systems use one year, some use two years, and some adjust the moving average period based on the previous forecast accuracy. For this example we will use a moving average period of twelve months.
2. Calculate the old average through June of this year. Exponential smoothing approximates moving average periods by using an alpha factor to weight the old average and the new average. The alpha factor is the number 2 divided by (1 plus the number of periods). Figure A7.2 shows how to calculate the old average.
3. Select a moving average period for the trend. Some exponential smoothing systems use one year, others use six months. We will use a moving average period of six months.

Figure A7.2   CALCULATE THE OLD AVERAGE

Last year	New average	=	Current period	×	Alpha	+	Old average	×	(1 − Alpha)
Jan	6	=	6	×	1	+	0	×	0
Feb	144	=	212	×	0.67	+	6	×	0.33
Mar	261	=	378	×	0.50	+	144	×	0.50
Apr	208	=	129	×	0.40	+	261	×	0.60
May	188	=	163	×	0.33	+	208	×	0.67
June	161	=	96	×	0.29	+	188	×	0.71
July	163	=	167	×	0.25	+	161	×	0.75
Aug	162	=	159	×	0.22	+	163	×	0.78
Sept	170	=	201	×	0.20	+	162	×	0.80
Oct	167	=	153	×	0.18	+	170	×	0.82
Nov	142	=	76	×	0.17	+	167	×	0.83
Dec	126	=	30	×	0.15	+	142	×	0.85
This year									
Jan	118	=	72	×	0.15	+	126	×	0.85
Feb	114	=	90	×	0.15	+	118	×	0.85
Mar	113	=	108	×	0.15	+	114	×	0.85
Apr	116	=	134	×	0.15	+	113	×	0.85
May	113	=	92	×	0.15	+	116	×	0.85
June	117	=	137	×	0.15	+	113	×	0.85

4. Calculate the old average trend through June of this year. The new average trend measures the average change in the new average. For example, the new average for January of last year was 6. The new average for February of last year was 144. The new average trend from January to February was 138 up. The new average for March was 261. The trend from February to March was 117 up. The new average trend for January to February and February to March was 123 up. Figure A7.3 shows the calculations for the old average trend through June of this year.

5. Select a seasonal index for broiler pans. Some exponential smoothing systems use a group index for all housewares, some use seasonal indexes based on prior years. We will use a

Figure A7.3   CALCULATE THE OLD AVERAGE TREND
( ) ARE NEGATIVE NUMBERS

Last year	New average	=	Current period	×	Alpha	+	Old average	×	(1 − Alpha)
Jan	0								
Feb	138	=	138	×	1	+	0	×	0
Mar	123	=	117	×	0.67	+	138	×	0.33
Apr	35	=	(53)	×	0.50	+	123	×	0.50
May	13	=	(20)	×	0.40	+	35	×	0.60
June	1	=	(17)	×	0.33	+	13	×	0.67
July	1	=	2	×	0.29	+	1	×	0.71
Aug	0	=	(1)	×	0.29	+	1	×	0.71
Sept	2	=	8	×	0.29	+	0	×	0.71
Oct	1	=	(3)	×	0.29	+	2	×	0.71
Nov	(7)	=	(25)	×	0.29	+	1	×	0.71
Dec	(10)	=	(16)	×	0.29	+	(7)	×	0.71
This year									
Jan	(9)	=	(8)	×	0.29	+	(10)	×	0.71
Feb	(8)	=	(4)	×	0.29	+	(9)	×	0.71
Mar	(6)	=	(1)	×	0.29	+	(8)	×	0.71
Apr	(4)	=	3	×	0.29	+	(6)	×	0.71
May	(4)	=	(3)	×	0.29	+	(4)	×	0.71
June	(2)	=	4	×	0.29	+	(4)	×	0.71

seasonal index based on the last calendar year. Figure A7.4 shows how to calculate the seasonal index.

6. Calculate the new average for July.

$$\frac{\text{New}}{\text{Average}} = \frac{\text{Current}}{\text{period}} \times \text{Alpha} + \frac{\text{Old}}{\text{average}} \times (1\text{-Alpha})$$

July,
this year   117   =   120   × 0.15 +   117   ×   0.85

7. Calculate the new average trend for July.

$$\frac{\text{New}}{\text{Average}} = \frac{\text{Current}}{\text{period}} \times \text{Alpha} + \frac{\text{Old}}{\text{average}} \times (1\text{-Alpha})$$

July,
this year   (2)   =   0   × 0.15 +   (2)   ×   0.85

Figure A7.4 BROILER PAN SEASONAL INDEX

Last year	Unit demand	Percentage of year
Jan	6	0.3
Feb	212	12.0
Mar	378	21.4
Apr	129	7.3
May	163	9.2
June	96	5.4
July	167	9.4
Aug	159	9.0
Sept	201	11.4
Oct	153	8.6
Nov	76	4.3
Dec	30	1.7
Total	1770	100.0

8. Calculate the expected new average for August, September, and October.

August new average   = July new average + (1 × new trend)
                    = 117 + (1 × (–2))
                    = 115

September new average = July new average + (2 × new trend)
                    = 117 + (2 × (–2))
                    = 113

October new average   = July new average + (3 × new trend)
                    = 117 + (3 × (–2))
                    = 111

9. Calculate the forecast for August, September, and October.
   Forecast = expected new average × 12 × seasonality index
   August   124 = 115 × 12 × .0.09
   September 155 = 113 × 12 × 0.114
   October   115 = 111 × 12 × 0.086
   Total   384

10. Compare the exponential smoothing forecast to the actual demand and to focus forecasting (fig. A7.5). Focus forecasting outperforms exponential smoothing. In any group of items

over a period of time it always outperforms exponential smoothing.

Figure A7.5 COMPARISON OF EXPONENTIAL SMOOTHING AND FOCUS FORE-CASTING FOR A BROILER PAN DEMAND, IN UNITS

	Exponential smoothing	Actual demand	Focus forecasting
August	124	151	116
September	155	86	116
October	115	113	116
Total	384	350	348

Focus forecasting is easier to explain than exponential smoothing. For example, here focus forecasting used the strategy "Whatever we sold in the past three months is probably what we will sell in the next three months." That's easier to explain than old averages, alphas, trends, seasonalities, and expected new averages.

Appendix 8

SALES FILTERS

Errors sometimes create bad demand in our records. Many forecasting systems use sales filters to screen out bad demand. These sales filters create as many problems as they solve. Figure A8.1 shows an example of bad demand in July. We want a range that tells us whether this year's July demand of 920 units is good or bad. Here's how we find that range.

Figure A8.1 PRODUCT X DEMAND IN UNITS

	Jan	Feb	Mar	Apr	May	June	July	Aug	Sept	Oct	Nov	Dec
Last year	6	212	378	129	163	96	167	159	201	153	76	30
This year	72	90	108	134	92	137	920					

1. Calculate the total absolute forecast error for the past six months.

	Forecast	Actual	Difference
Jan	4	72	68
Feb	157	90	67
Mar	272	108	164
Apr	94	134	40
May	124	92	32
June	71	137	66
Total absolute forecast error			437

2. Calculate the average absolute forecast error for the past six
    months.
    437 ÷ 6 = 73 average absolute forecast error
3. Calculate a forecast for July.
    130 forecast for July
4. Calculate a range that tells us whether this year's July demand
    of 920 units is good or bad. The average absolute forecast error
    is about 80 percent of one standard deviation. A 3.75 average
    absolute forecast error is the range above and below the
    forecast that we will accept as good actual demand.

High range = July forecast + 3.75 × average absolute forecast
             error
           = 130 + 3.75 × 73
           = 404 units
Low range = July forecast −3.75 × average absolute forecast
             error
           = 130 −3.75 × 73
           = 0

5. Do not accept this year's July demand of 920 units because it
    falls outside 0 and 404 units.
6. Change the recorded July demand to 404 units!

This is the basic logic for sales filters. Whenever demand exceeds
the acceptable range it is changed.

If product X is a broiler pan this practice probably won't hurt, but
if it is a pump in a flood, a gasoline can in an energy crisis, or a flag
in the bicentennial year, the sales filter would be a disaster. The
simple truth is that in any year, almost any item can have
exceptional demand. Sales filters create as many problems as they
solve.

A sales filter can flag items for review. At the end of chapter 2 is a
description of a simple sales filter which does not change item
demand but does signal the buyer to review items with unusual
demand.

Appendix 9

PROBABILITY DISTRIBUTIONS
LINE-FILL RATE VERSUS MONTHS ON HAND

It takes a much greater increase in months on hand to move from a line-fill rate of 87 percent to a rate of 90 percent than it does to move from a rate of 85 percent to a rate of 88 percent.

In a normal distribution of random occurrences, standard deviations measure the range of possible occurrences. These statistics of chance have been proven over the years. As long as there are enough occurrences and a long enough period of time, these statistics of chance are laws just like the law of gravity.

The key to a normal distribution is a standard deviation. A standard deviation measures the distribution of occurrences about an average value. A standard deviation is usually shown as the Greek letter sigma σ. Figure A9.1 shows a normal distribution and standard deviations.

Figure A9.1   A NORMAL DISTRIBUTION

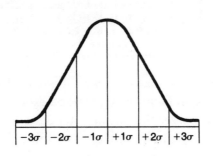

| -3σ | -2σ | -1σ | $+1\sigma$ | $+2\sigma$ | $+3\sigma$ |

Normal distributions have certain rules:

1. Within one standard deviation ($\pm\sigma$) of the average value 68 percent of all the outcomes occur.
2. Within two standard deviations ($\pm 2\sigma$) of the average value 95 percent of all the outcomes occur.
3. Within three standard deviations ($\pm 3\sigma$) of the average value 99.7 percent of all the outcomes occur.

Notice that the first standard deviations from the average covered 68 percent of all the outcomes, but the next standard deviation covered only another 27 percent of the outcomes and the third covered only another 4.7 percent of the outcomes.

Figure A9.2 LINE-FILL RATE % VS MONTHS ON HAND

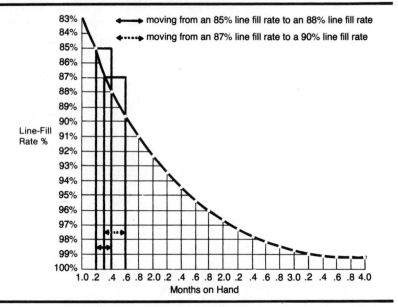

Months on hand are related to line-fill rate by the laws of chance. The more months on hand there are, the fewer customer lines will be canceled. You can see that the first standard deviation of months on hand covers the greatest number of customer line cancellations. The first increase in safety stock from none to some has the greatest impact on line-fill rate. For this reason it takes a much greater increase in months on hand to move from an 87 percent line-fill rate to a 90 percent line-fill rate than it takes to move from an 85 percent rate to an 88 percent rate.

Appendix 10

SIMULATING GOALS

The inventory manager must set short-term realistic inventory goals. The general manager usually has some goals in mind already, such as achieving a customer service level of 98 percent with an inventory turnover of 6. The inventory manager may find that in the past ten years the only time the company hit both goals on the same date was the year the southern warehouse burned down during a mail strike!

The size of the inventory investment has a cost. The handling of frequent orders has a cost. The loss of customer sales has a cost. Thus the overall inventory goal should be to find the combination of these three costs that produces the least possible total cost. The inventory manager makes two decisions that determine the proportion of these costs. He decides review time and safety stock time. Review time is the time between buying decisions. The longer the review time, the higher the inventory investment and the lower the frequency of ordering. The safety stock time is the time the company can wait for a late vendor order without running out of stock. This same safety stock time covers low forecast errors. The longer the safety stock time, the higher the inventory investment and the higher the customer service level.

The computer can simulate the impact of different review and safety stock times on inventory investment and customer service using last year's demand data. Then the inventory manager can use these simulations to set realistic short-term goals.

Suppose the existing review time is 1.0 months. Let's simulate various safety stock time settings and their impact on inventory turnover and customer service. The computer counts line cancellations and calculates a line-fill rate at various safety stock times. Then it divides the year's demand in dollars by the average dollars on hand to calculate the inventory turnover at these safety stock settings. The results appear in figure A10.1.

Figure A10.1   SIMULATION OF SAFETY STOCK SETTINGS ON TURNOVER AND LINE-FILL RATE

	Review Time = 1.0 Months				
Safety stock time (months)	0	0.5	1.0	1.5	2.0
Inventory turns per year	22	10	7	5	4
Customer service (Line-fill rate)	73%	87%	88%	91%	93%

This figure shows the company's current ability to achieve a given customer service level with a given inventory turnover. The relationship is based on the company's current inventory management system, its current suppliers, current product mix, and existing work load. The inventory manager may also simulate different review times; changes in review time will affect the company's work load. If the inventory manager does not change review time, this simulation gives the only realistic short-term goals.

Appendix 11

DERIVATION OF THE CLASSIC FORMULA FOR ECONOMIC ORDER QUANTITY

The larger the order quantity, the higher the average inventory and the resulting inventory carrying cost. The smaller the order quantity, the higher the frequency of ordering and the resulting inventory ordering cost. The total of these two costs as a function of order size is expressed in the following formula:

$$T_c = \frac{SC_o}{Q} + \frac{QC_c}{2}$$

where T_c = total cost for one year
  S = sales in units for one year
  C_o = cost of ordering one order
  C_c = cost of carrying one unit for a year
  Q = order quantity

Now to find the order quantity that minimizes the total cost, we first differentiate the total cost T_c with respect to the order quantity Q.

$$\frac{DT_c}{DQ} = -SC_o Q^{-2} + 0.5C_c$$

The minimum point in the total cost T_c curve in relation to the order quantity Q is then the point at which

234

$$\frac{DT_c}{DQ} = 0$$

So

$$0 = -SC_oQ^{-2} + 0.5C_c$$

$$SC_oQ^{-2} = 0.5C_c$$

$$Q^{-2} = \frac{0.5C_c}{SC_o}$$

$$Q^2 = \frac{SC_o}{0.5C_c}$$

$$Q = \sqrt{\frac{SC_o}{0.5C_c}}$$

This is the derivation of the classic economic order quantity formula.

DERIVATION OF THE FORMULA FOR THE NEW
CONCEPT OF EOQ

Order frequency determines work and inventory size. The more frequently we order, the more work we have and the less inventory we carry. The less frequently we order, the less work and the more inventory we carry.

For a group of items or suppliers we have a total inventory cost that can be expressed as the number of lines of work involved in ordering the group of items or suppliers plus some percentage x of the inventory investment in the group of items or suppliers. The total cost can then be expressed as some total lines of work.

$$
\begin{aligned}
\text{Let } T_L &= \text{ total lines of work} \\
P_F &= \text{ ordering frequency} \\
N_i &= \text{ number of items} \\
A &= \text{ annual sales in dollars} \\
x &= \text{ inventory carrying cost percentage in terms of lines} \\
P_F N_i &= \text{ ordering cost in terms of lines of work} \\
0.5 P_F^{-1} A x &= \text{ carrying cost in terms of lines of work} \\
T_L &= P_F N_i + 0.5 P_F^{-1} A x
\end{aligned}
$$

The total lines of work can be differentiated with respect to the order frequency.

$$
\frac{DT_L}{DP_F} = N_i - 0.5 P_F^{-2} A x
$$

Then the minimum total lines of work order frequency occurs at the point where the change in the total lines of work with respect to the order frequency is zero, that is where

$$\frac{DT_L}{DP_F} = 0$$

Thus

$$0 = N_i - 0.5 P_F^{-2} A x$$

$$P_F^{-2} = \frac{N_i}{0.5 A x}$$

$$P_F = \sqrt{\frac{0.5 A x}{N_i}}$$

This is the formula for calculating the order frequency that minimizes total costs in terms of lines of work.

Appendix 13

ITEM DEMAND HISTORIES

Here are the twenty-four month demand histories for twenty different items. They are real demand histories, so don't look for any particular beauty in their demand patterns. Some items are seasonal. Some exhibit trends. Most exhibit noise. See whether you can create a focus forecasting system that produces usable results. See whether your system can beat people's guesses of this year's August through January actual demand.

	Feb	Mar	Apr	May	June	July	Aug	Sept	Oct	Nov	Dec	Jan
Transmission Sealer												
This year	16	35	41	34	30	51	23	16	23	12	17	31
Last year	25	21	2	27	18	44	12	13	2	1	6	26
Screwdriver												
This year	20	31	9	16	13	21	25	23	6	4	12	70
Last year	14	6	14	12	11	51	6	35	30	12	17	26
Wood Heater												
This year	43	60	17	18	29	15	5	100	154	128	145	139
Last year	51	56	41	10	5	13	44	117	135	151	183	203
Thumbtacks												
This year	150	220	130	130	210	160	220	240	120	240	330	270
Last year	240	120	130	200	220	90	220	140	290	200	560	170
Tape Recorder												
This year	3	11	3	0	0	7	11	2	2	12	17	4
Last year	14	23	9	8	7	3	6	12	8	4	9	19
Choke Chain												
This year	52	57	24	33	70	38	77	33	41	38	31	25
Last year	45	69	66	29	61	28	54	49	48	27	7	48

	Feb	Mar	Apr	May	June	July	Aug	Sept	Oct	Nov	Dec	Jan
Rifle Sling												
This year	1	0	6	1	2	7	17	11	5	24	13	0
Last year	4	12	0	1	13	0	0	14	7	8	2	10
Drill												
This year	25	24	20	26	13	58	44	18	29	30	31	33
Last year	7	0	0	3	30	16	12	11	16	14	23	19
White Paint												
This year	24	76	130	112	212	199	124	122	52	14	21	16
Last year	12	30	136	181	178	151	132	71	123	66	26	24
Dandelion Digger												
This year	51	163	568	340	107	46	25	4	10	0	0	15
Last year	37	113	205	488	142	48	38	18	0	0	0	10
Motor Oil												
This year	1040	736	918	593	700	590	550	1221	867	1027	789	877
Last year	841	894	698	774	997	650	718	705	1040	1458	1112	1737
Shutter												
This year	274	319	201	342	373	218	227	109	352	739	162	250
Last year	154	213	231	421	179	209	452	209	379	324	334	327
Corn Broom												
This year	480	780	648	468	360	384	468	504	660	612	552	756
Last year	660	528	396	408	408	300	420	432	420	360	596	504
Furnace Filter												
This year	4332	1704	1008	612	708	516	780	2664	3324	3552	3192	5004
Last year	2052	1392	864	792	828	480	540	3288	4344	2088	2232	2352
Mousetrap												
This year	1224	1836	1440	1188	2520	1656	2196	1116	828	900	936	1044
Last year	720	900	972	1188	2016	2232	1764	828	1368	1476	720	792
Gas Cylinder												
This year	7452	1680	852	1356	2292	2424	2592	1872	1644	2040	3012	13224
Last year	2016	984	624	1080	1860	1704	1452	1584	1188	1200	1800	4260
Light Bulb												
This year	2928	3264	3888	4272	4320	4176	4416	3024	4416	3456	3840	3216
Last year	4752	4368	4944	4368	4128	5184	5664	4416	4416	4368	4224	4368
Nails												
This year	1800	2500	1800	2300	1550	600	850	500	1500	1800	1500	950
Last year	2600	2500	1750	2350	1250	1300	1200	1350	1400	1950	2000	1550
Manure Fork												
This year	2	25	29	20	13	13	29	21	12	19	12	15
Last year	8	28	41	31	49	39	15	15	28	39	19	11
Weed Cutter												
This year	35	89	94	282	555	648	572	230	41	19	6	0
Last year	38	59	78	184	668	579	476	274	44	41	6	1

Appendix 14

REVIEW QUESTIONS

Chapter 1

1. Why is inventory management never right? Why is inventory management such an important part of the management of almost any business?
2. Many people use the KISS (keep it simple stupid) approach for the development of new systems. The systems in this book are simple. They are based on three principles. What are they?
3. Focus forecasting is a new approach to forecasting item demand. Describe how it works.
4. Focus forecasting results can be measured in three ways. What are they?
5. One of the earliest approaches to computerized forecasting was the recipe approach. List the four ingredients in the recipe approach and describe each of them.
6. How are cycles like seasonality? How do they differ?
7. Why is the recipe approach more accurate in forecasting high-volume demand?
8. What is exponential smoothing? Why was it originally developed?
9. What is computer simulation? What was it originally used for? How does focus forecasting use computer simulation?
10. What is the difference between a statistical approach to forecasting and a simulation approach to forecasting? Why is

the simulation approach to forecasting more appropriate for a company with thousands of inventory items?

11. Business itself is complicated, yet there are many complicated approaches to solving inventory management problems. What might be one reason that some people develop complicated systems?

Chapter 2

1. This chapter illustrates patterns of demand for different items. The focus forecasting example in this chapter used two simple approaches. Method A was, Whatever the demand was in the past three months will probably be the demand in the next three months. Method B was, Whatever percentage increase or decrease we had over last year in the past three months will probably be the percentage increase or decrease we will have over last year in the next three months. Use these two simple strategies in a focus forecasting approach to forecast three of the items in appendix 13.

2. Here are three more simple forecasting strategies. (1) The next three months' demand will be three times last month's demand. (2) The next three months' demand will be one quarter of the past twelve months' demand. (3) The next three months' demand will be last year's demand for the next three months. Use these simple strategies, or any others that you choose, with the focus forecasting approach to forecast one of the items in appendix 13.

3. Cover up the actual demand for the last six months of this year for the item that you selected from appendix 13. Ask someone to guess the demand for the last six months for this item. Compare the actual demand against your focus forecasting results. Compare the actual demand against the guess. If the guess of this item's demand was more accurate than your focus forecasting system ask the person who made the guess to explain to you the procedure he used to forecast the item. If the guess was less accurate than your focus forecasting system, skip review question 4.

4. Incorporate the forecasting strategy the person used to make a guess of the item demand into your focus forecasting strategy. Repeat questions 3 and 4 until your focus forecasting approach outperforms guesses of future demand.

5. Focus forecasting does not use seasonality curves to forecast new item demand. Why does using seasonality curves to forecast new item demand cause more problems than it cures?

6. What is the fly swatter philosophy? Identify ten different items for which the fly swatter philosophy might improve forecasting accuracy.

7. Computerized forecasting systems often run into trouble when they encounter extraordinary demand. Identify three causes of extraordinary demand.

8. Describe a method for selecting extraordinary item demand for buyer review.

9. Why is it important to get the strategies for your focus forecasting system from the people who will use the system?

10. The keys to maintaining the trust of those who use your focus forecasting system are simplicity and transparency. What is meant by simplicity? What is meant by transparency?

11. For the most part it is better not to include exotic forecasting, strategies like exponential smoothing, cumulative line ratio forecasting, or least squares regression in your focus forecasting system, even though these forecasting approaches have produced usable degrees of accuracy for many companies. Why is it not advisable to use them in your focus forecasting strategy?

Chapter 3

1. What is material requirements planning? How do manufacturing companies use it?

2. What is finished goods inventory management? How do manufacturers use it? How do wholesalers and retailers use it?

3. How are the jobs of the following people similar in both wholesaling and manufacturing companies? How are their jobs different? inventory manager, sales manager, controller,

operations manager, buyer, expediter, and the general manager.

4. What are the three basic objectives in finished goods inventory management?

5. How do companies measure customer service?

6. How do companies measure the size of their inventory?

7. What are some of the cost elements of ordering costs? What are some of the elements of inventory carrying costs?

8. What are the six basic steps in this finished goods inventory management system?

9. The inventory manager is well-versed in techniques for forecasting total dollar demand. Why does the inventory manager use the controller's sales budget in preparing the total inventory plan?

10. The inventory manager looks at historic performance to plan the total inventory levels. How does he convert the controller's sales budget into planned total dollars of inventory for each month of the coming budget period?

11. How does this finished goods inventory management system marry the financial plans of the company to the buying of individual items?

12. In this finished goods inventory management system the buyers change less than 8 percent of the computer's suggested purchases. Why don't they change more of these suggested purchases?

13. What is the primary objective of the expediting department? What is the only way they can accomplish this objective?

14. The controller measures the inventory performance in the financial statement. What other measurements does the inventory manager make of the total inventory management system's performance?

15. The inventory level on June 30 is expected to be $10 million. The projected sales from the controller's sales budget for July are $6 million. The planned inventory for the end of July is $11 million. Assuming that it takes one month to replenish inventory, how many dollars must the inventory manager

purchase in June to meet the total (dollar) planned inventory level for the end of July?

Chapter 4

1. What are some arguments for increasing the size of the inventory?

2. Why is it important that the inventory manager's inventory plan become part of the company's operating budget?

3. The inventory manager wants a realistic sales budget from the controller. Why do controllers tend to underestimate budgeted sales?

4. Why and how does the inventory manager convert the sales budget to sales at cost?

5. What is line-fill rate? What is dollar-fill rate?

6. Why should line-fill rate goals be higher in the company that backorders than in a company that does not?

7. Discuss the advantages and disadvantages of the classic turnover calculation versus the months-of-supply measure of turnover.

8. If having 2.0 months on hand at the end of June produces a 90 percent fill rate in July, can the inventory manager assume that having 2.0 months on hand at the end of July will produce a 90 percent fill rate in August?

9. This chapter mentions ten considerations that the inventory manager should look at in setting line-fill rate goals. Name seven of them.

10. Why is it better to plan high line-fill rates at the beginning of a season and low line-fill rates at the end of a season?

11. At the end of July the total inventory on hand is $11 million. The sales budget at cost for August is $7 million. The sales budget at cost for September is $5 million. Compute the months on hand at the end of July.

12. There are four columns on the inventory plan—planned inventory, budgeted sales at cost, planned months on hand, and line-fill rate. What is the basic input for calculating each of these items?

13. The controller, the sales manager, and the operations manager

look at the inventory plan from different perspectives. Describe what each manager is looking for in the inventory plan.

14. The operations manager has the largest number of people in the company, while the inventory manager has control of one of the largest assets in the company. How should the financial report be used to measure each of their performances?

15. The September 30 planned inventory on hand is $14,080,000. The budgeted sales at cost for September are $6,300,000. The planned inventory for August 31 is $11,240,000. Assuming a one-month inventory replenishment cycle how many dollars should the inventory manager purchase in August to meet the September 30 planned inventory goal?

16. A safety stock setting of 0.4 to 1.4 months generates purchases of $7,609,000. A safety stock setting of 0.8 to 1.8 months generates purchases of $8,219,000. What should the safety stock setting be to generate $8 million worth of purchases?

17. What is the dilemma that faces the inventory manager in planning the inventory?

Chapter 5

1. When the available inventory falls below the reorder point, most inventory management systems order an economic order quantity. How is this procedure different in this finished goods inventory management system?

2. How do you convert the reorder point time to a reorder point quantity?

3. What are the three lengths of time included in reorder point time?

4. Name each of these elements of time: (a) the time that passes between each buying decision, (b) the time it takes goods to get on the shelf after a buying decision, and (c) the time that a company can wait for a late vendor's order without running out of stock.

5. How does review time affect the amount of work in the company and the size of the company's inventory? How does review time affect a company's ability to meet minimum freight restrictions and maximum quantity discounts?

6. Why is consistent lead time more important than its length?
7. Many inventory management systems determine the size of safety stock based on the amount of forecast error. How does this procedure affect the customer service on the best-selling items in the company?
8. The three elements of reorder point time are review time, safety stock time, and lead time. What portion of each of these elements of time determines the average on-hand inventory balance?
9. The company carries a two-inch paintbrush in its inventory. The company reviews this paintbrush for a buying decision twice a month. On the average it takes about one month for a paintbrush to arrive in inventory after a purchase order is issued. The company carries one month's safety stock to cover inconsistencies in supplier lead times and errors in forecasting. The company sells about 100 of these paintbrushes every month. At the end of any month what is the average quantity for the reorder point, the on-hand inventory balance, and the on-order balance? On the average, how many paintbrushes does the company buy each time it reviews the paintbrushes for a buying decision?
10. On any item that the company purchases the expected line-fill rate can be influenced by a number of factors. In this chapter you have seen eight of these factors. List four of them.
11. In figure 5.10 the current on hand at the beginning of July was 50 lawn mowers. The forecasted demand for July and August was 1800 lawn mowers. Yet the finished goods inventory management system purchased only 300 lawn mowers in July. This is an example of end-of-season reorder point. Why did the finished goods inventory management system buy only 300 lawn mowers?
12. In figure 5.10 what would happen if the lead time was only one month? How many lawn mowers would the finished goods inventory management system purchase?
13. In this finished goods inventory management system the inventory manager controls every element of time in the reorder point time. How does he control review time? How

does he control safety stock time? How does he control lead time?

Chapter 6

1. The reorder point time consists of safety stock time, review time and lead time. The new concept of EOQ presented in this chapter is based on review time. What is the definition of the ideal review time?
2. With this concept of economic order quantities the inventory manager can change the parameters in the computer simulation to achieve four different objectives. What are they?
3. An order quantity is the number of units to be bought or made when an inventory needs replenishment. What is an economic order quantity?
4. Why do inventory managers want small order quantities? Why do production people and warehousing people prefer larger order quantities?
5. List five elements of ordering costs in a manufacturing company.
6. List six elements of ordering costs in a wholesaling company.
7. List six elements of inventory carrying costs in wholesaling companies and in manufacturing companies.
8. A classic economic order quantity formula requires three estimates to arrive at an economic order quantity for an item. What are these three estimates?
9. One of the major limitations in the classic economic order quantity formula is that it assumes costs are linear. What is meant by the term linear? Why are costs in manufacturing and in wholesaling not linear?
10. Many wholesalers and retailers are using item economic order quantity formulas. Why is it surprising that they use these formulas? What sort of economic order quantity formulas should they use?
11. This new concept of EOQ is based on four observations of most companies. What are the four observations?
12. This new concept of economic order quantities requires no estimates of ordering costs or inventory carrying costs, yet it

reduces both ordering and inventory carrying costs. How does the inventory manager measure the reduction in order costs and the reduction in inventory carrying costs?

13. In this economic order quantity formula, how is inventory carrying cost expressed?

14. A company has two vendors, vendor C and vendor D. Here is a schedule of the annual volume with each vendor:

	Vendor C	Vendor D
Number of items	2	10
Review frequency	12	12
Annual volume	$360,000	$60,000
Minimum freight restrictions	none	none

(a) Reduce the lines of work in the company and the company's average inventory investment by the same percentage. (b) Minimize the lines of work in the company while maintaining the company's average inventory investment. (c) Minimize the company's average inventory investment while maintaining the lines of work in the company.

15. What are minimum freight restrictions? What are maximum quantity discounts? How do they affect gross profit?

16. Calculate the ideal review frequency for vendor C and vendor D assuming that vendor C has a $40,000 minimum freight restriction.

17. This new economic review frequency formula obviously has application for wholesalers and retailers. Discuss whether this concept is applicable to manufacturing companies.

Chapter 7

1. What is a distribution requirements plan? How is it similar to a material requirements plan? How is it different?

2. The reorder point time is made up of safety stock time, review time, and lead time. Which of these three times can the inventory manager control through distribution requirements planning?

3. List three reasons for an inventory manager to set up a distribution requirements plan with major suppliers.

4. Many companies give their suppliers a schedule of their future requirements, but these schedules are not distribution requirements plans. What are the four points of agreement in a distribution requirements plan?

5. Both manufacturers and wholesalers invest in safety stock. A distribution requirements plan can eliminate one consideration in the safety stock strategy for both the manufacturer and the wholesaler. What elements of safety stock does distribution requirements planning eliminate?

6. Inconsistencies in vendor lead time cause the inventory manager two problems. What are they? How does distribution requirements planning help to eliminate these two problems?

7. As part of a distribution requirements plan the inventory manager agrees to make firm purchase order commitments far enough into the future so that the supplier can guarantee delivery. Discuss the advantages and disadvantages to the inventory manager of this firm purchase order commitment.

8. When the inventory manager starts up a distribution requirements plan with a major supplier he issues two purchase orders. What is the purpose of each of these purchase orders?

9. The manufacturer has the distribution requirements plan from the inventory manager tied into material requirements planning. What does the manufacturer do when the inventory manager increases the forecast of future demand for an item? What does the manufacturer do when the inventory manager decreases the forecast of future demand for an item?

10. Look at figure 7.2. The raw material manager for the manufacturer has built in one month's safety stock on broiler pan handles, but not by increasing the size of the on-hand inventory. How did he do it?

11. A manufacturer's raw material requirements plan can become a distribution requirements plan. Under what circumstances can this happen?

12. What is the difference between independent and dependent demand? Why should a company always try to use distribution requirements planning rather than a forecast of dependent demand? How does distribution requirements planning

improve turnover and customer service for the inventory manager? For the manufacturer?

Chapter 8

1. Give three reasons why the expediter's job theoretically should not exist.
2. Give five reasons why the expediter's job does exist.
3. Some expediters just communicate order numbers to vendors and write down shipping dates. How does a professional expediting staff insure the delivery of needed goods?
4. In this chapter we have seen ten things that the expediter can learn about the vendor's company to understand the company he is dealing with. List six of these things.
5. List five of the things an expediter can learn about the vendor's customer service representative. Why is it important for the expediter to learn these things?
6. What is the expediter's only consistent means of insuring the delivery of needed goods?
7. The expediting system in this chapter answers two questions. What are they?
8. The expediter's job does not end when the vendor ships the goods. When does it end?
9. What are priority dates? How are they calculated?
10. What are fixed priority dates? Does distribution requirements planning use calculated priority dates or fixed priority dates?
11. What is the float report? How is the information for the float report gathered?
12. List four receiving department priorities that may override priority dates.
13. In some companies the expediting department answers calls from customers on expected delivery. Give two reasons for having a separate customer service department to answer customer inquiries on delivery.
14. How should the expediting department communicate the status of delivery of needed goods to the customer service department?

Chapter 9

1. Measuring performance is the most important step in the finished goods inventory management system. The inventory manager measures the buyers' performance in three key areas. What are they?
2. The three key measures of buyer performance are like the three legs of a stool. If one is measured without the other two, the stool will fall over. Give an example of what happens when only one of the three key areas is measured.
3. What are the four major reasons for failing to ship a customer's order for an item?
4. Why would an operations department ever fail to ship a customer's order when the goods were on hand?
5. There are five major jobs in a warehouse department. How does an operations manager usually rank these five jobs in priority sequence? Why does he use this particular ranking?
6. When does the expediting system consider vendor lead time to be the cause of a customer cancellation?
7. Sometimes the inventory manager must cancel customer orders simply because he did not order enough. What are the three major reasons for failing to order enough?
8. List the action that the inventory manager should take in each of the following situations: (a) Goods are available but weren't shipped. (b) Goods are at the receiving dock but not put away for order filling. (c) The vendor or the production plant has failed to deliver on time. (d) The buyer didn't order enough.
9. What are the only two ways to improve turnover?
10. This chapter mentions two ways the inventory manager can provide information that can be used to increase sales. What are they?
11. The inventory manager must identify significant inventory excesses. What two characteristics determine that an inventory excess is significant?
12. The buyer can take four steps to reduce inventory excesses. What are they? List these steps for reduction of inventory excesses in the order of least cost to the company.

13. The inventory manager measures the buyers on return on inventory. Discuss how each of the following affects total return on inventory: (a) the initial gross profit percentage, (b) the discount taken on sales to reduce inventory excesses, (c) the initial inventory turnover, (d) the turnover lost through customer cancellations.

14. Why does the inventory manager measure the initial gross profit percentage of goods on order? of goods on hand?

15. A buyer's final gross profit on sales is 16 percent and his return on inventory is 48 percent. What is the buyer's inventory turnover?

16. The inventory manager measures primarily the buyer's performance. Other departments in the company affect inventory turnover and customer service, and the inventory manager measures the performance of these other departments. Which departments are they? How does the inventory manager measure them?

Chapter 10

1. There are many sophisticated and usable approaches to solving inventory problems. Before a company uses any of these approaches it should know the answer to one question. What is that question?

2. Identify the seven steps in creating new systems. Why are the first step and the seventh step so important?

3. Identify five areas in which a company should measure performance in its existing systems.

4. Why is it valuable to measure performance in a company's existing system even if no new systems are planned?

5. Describe the difference between ideal objectives, reasonable objectives, and realistic objectives.

6. What are the three major objectives of a finished goods inventory management system?

7. Sometimes a company finds the causes of a problem and decides that it does not need a new system. Its analysis of the causes of a problem may lead the company to make two changes other than creating new systems. What are they?

8. Almost every book on systems says that users must be involved in the creation of new systems. How can you involve users in the creation of new systems?
9. Why is it important to test new systems on past data before installing them?
10. List three specific areas that can be measured after the installation of a new finished goods inventory management system.
11. Why should a company avoid hasty changes in its existing finished goods inventory management system?

Chapter 11

1. What three qualifications should a company look for in an inventory manager? Why?
2. Explain what is meant by each of these qualifications.
3. How should you evaluate the qualifications of an inventory manager who is already inside the company?
4. The computer can provide tremendous assistance in setting realistic inventory goals. How can you use the computer to begin to set realistic inventory goals for your company?
5. The company president has realistic inventory goals. The inventory manager has realistic inventory goals. Whose goals should be used in the short term and whose should be used in the long term?
6. The first step in long-range planning is to understand the company's existing situation. What four questions help the inventory manager to understand the company's current inventory management situation? Who should answer these questions?
7. The long-range plan in this chapter has five parts: inventory goals, objectives and procedures, people, costs, and benefits. The goals come from the inventory manager and the company president. Discuss how the other four parts of the inventory plan are related to the company's current inventory situation. For example, one concern in this example of a company was the loss of computer records. The first objective in the long-

range plan was to keep our existing customer service and inventory turnover.

8. The example of a long-range plan presented in this chapter breaks costs into development costs and operating costs. What is the difference between development costs and operating costs?

9. In the sample long-range plan how are the benefits related to the inventory goals?

10. Discuss how the existing inventory management situation in a company can suggest objectives and procedures for improving inventory management performance.

Chapter 12

1. In some companies data processing is a line function with its own objectives. To create new systems that work the data processing staff must become part of the company team. What six things can make data processing people part of the company team?

2. To become part of the company team the data processing staff must specialize in the department they service. They must become business persons in addition to being technicians. How do some of our businesses and colleges encourage data processing people to remain technicians?

3. How does having a member of the data processing staff specialize in a department in the company make him more a part of the company team?

4. What are some examples of business language that data processing people have problems with? What are some examples of data processing language that business people have problems with? How can this language problem be overcome?

5. Give three examples of procedures for teaching data processing people about business and business people about data processing.

6. Three ways to make data processing people more a part of the company are (a) let the data processing people specialize in the area of business they are servicing, (b) teach the data processing people about business and teach the business

people about data processing, and (c) share the rewards of successful systems with the people in data processing. Give examples of three ways that a company can let data processing share in the rewards of successful systems.

7. It has been said that even if line managers never asked for another new system the data processing department would remain as busy as they are now. What would they be doing?

8. Some data processing departments spend too much time improving internal efficiency. Give examples of ten questions they ask regarding improving their own efficiency. Why are these the wrong questions? What must they do to make these the right questions?

9. What is a data processing steering committee? What is its purpose?

10. Some data processing departments use a form to evaluate systems service requests. What are the five parts of this form? What is the most important question on that form?

11. How are requests for new systems like requests for capital expenditures?

Index